The Cambridge Eclipse Photography Guide describes everything you [illegible] to know to observe and photograph the forthcoming solar and [illegible] eclipses in the 1990s. It gives maps and descriptions of wher[e] [illegible] how to watch all these eclipses. Particular attention is given to th[e] [illegible] solar eclipses that will be the most popular: an annular eclipse (w[hen] a ring of sunlight is visible around the moon) that crosses the Un[ited] States in 1994 and a total eclipse that passes over Europe in 1999. The authors give instructions and tips on how to watch the eclipses safely and how to photograph them with still and video cameras. The book is well illustrated with color and black-and-white eclipse photographs. Past eclipse expeditions are also described, encapsulating the excitement of those observers involved. This is an invaluable handbook for those who wish to witness one of nature's most spectacular events visible on earth.

The Cambridge Eclipse Photography Guide

THE CAMBRIDGE ECLIPSE PHOTOGRAPHY GUIDE

How and where to observe and photograph solar and lunar eclipses

JAY M. PASACHOFF
Field Memorial Professor of Astronomy and Director of the Hopkins Observatory, Williams College

MICHAEL A. COVINGTON
The University of Georgia

With tables and maps of the eclipses by
FRED ESPENAK
NASA/Goddard Space Flight Center

Published by the Press Syndicate of the University of Cambridge
The Pitt Building, Trumpington Street, Cambridge CB2 1RP
40 West 20th Street, New York, NY 10011-4211, USA
10 Stamford Road, Oakleigh, Melbourne 3166, Australia

First published 1993

Printed in Great Britain at the University Press, Cambridge

A catalogue record for this book is available from the British Library

Library of Congress cataloguing in publication data

Pasachoff, Jay M.
The Cambridge eclipse photography guide/Jay M. Pasachoff, Michael A. Covington; with tables and figures of the eclipses compiled by Fred Espenak.
p. cm.
"How and where to observe and photograph solar and lunar eclipses"
Includes index.
ISBN 0-521-45651-7 (pbk.)
1. Astronomical photography. 2. Solar eclipses. I. Pasachoff, Jay M. II. Covington, Michael A., 1957–
Astrophotography for the amateur. IV. Title.
QB121.C69 1993
523.7'8–dc20 93-13054 CIP

ISBN 0 521 45651 7 paperback

RO

Contents

ACKNOWLEDGMENTS

Jay M. Pasachoff is glad to thank his wife, Naomi, and his daughters, Eloise and Deborah, for their participation in numerous eclipse expeditions and for their friendship and support. He thanks Susan Kaufman for her assistance. He thanks the National Geographic Society in general and Mary G. Smith in particular for their support of many of his eclipse expeditions between 1970 and the present. He remembers fondly the participation of many Williams College students on his expeditions and thanks also his colleagues Bruce Miller and Bryce Babcock for their eclipse work. He cherishes the memory of Phil Schierer, a colleague who participated with him on many eclipse expeditions and with whom he spent much time discussing the subject of this book. He is grateful for scientific grants from the Committee on Research and Exploration of the National Geographic Society and from the National Science Foundation. In particular, the expeditions and data reduction benefited from NSF grants RII-8304403 and National Geographic grant 2547-82 for the 1983 eclipse; NSF grants ATM-9005194, USE-9050643, and AST-9014889, and National Geographic grant 433-70 for the 1991 eclipse; and NSF grant ATM-9207110 for the 1994 eclipse. Prof. Pasachoff is eternally in the debt of the late Prof. Donald H. Menzel of the Harvard College Observatory for introducing him to solar eclipses and eclipse expeditions.

Michael Covington wants to thank everyone who assisted with his earlier book, *Astrophotography for the Amateur*, from which some of this material is derived, as well as the staff of the University of Georgia library map room. He also thanks his wife Melody and daughters Cathy and Sharon, who lent much-needed moral support.

The authors thank Simon Mitton and his staff and colleagues at Cambridge University Press for their interest in this work and for their expertise in the production of this book. They also thank Fred Espenak for his participation. They thank Nancy P. Kutner for making the index and Deborah Pasachoff, Naomi Pasachoff, Dennis di Cicco, Kevin Reardon and Bonnie Schulkin for reading the proof.

INTRODUCTION

DARKNESS SWEEPS THE EARTH

Through a filter, we could see that the moon was starting to cover the sun. The solar eclipse had begun. We had not travelled thousands of miles to Hawaii and brought three tons of equipment in vain. Nobody was to jump from behind a palm tree and shout "April fool." The most exciting spectacle that is ever visible on earth would soon occur: a total eclipse of the sun. The new day would soon turn back to night.

Dozens of scientists and tens of thousands of tourists had come to mid-Pacific in 1991 to watch, study, and marvel over the spectacular phenomena that occur when the moon completely hides the sun. In this book, we will tell you how to observe and photograph this kind of eclipse as well as the other kinds that occur. Over the next decade, many solar eclipses – both total, in which the sun is entirely hidden, and annular, in which a ring of sunlight remains – will be visible from one place or another across the globe. Further, many lunar eclipses, in which the moon is partly or entirely hidden by the earth's shadow, will be widely visible. All are fun to see.

Between us, we have seen dozens of eclipses, and we are glad to share our experiences with you. Sometimes we will write in the voice of one of us; at other times, we will write together. We hope to convey both technique and spirit.

By two thousand five hundred years ago, the Babylonians had found that solar eclipses repeated in a pattern. Taken by themselves, the patterns in the sky followed by the sun and the moon do not mesh. But when the earth has gone around the sun exactly 19 times, measured by the number of times the path of the sun in the sky crosses that of the moon, the moon has gone around the earth exactly 223 times. As a result, the sun and moon come back to the same place in the sky at exactly the same time. In fact, even the moon's distance from the earth, which varies because the moon's orbit is not round, is the same after this interval, which is called the "saros." The Babylonians had discovered this saros interval of 18 years 11⅓ days. (Our leap years sometimes make the 11 days into 10 or 12.) The ⅓ day gives the

earth a chance to rotate an extra ⅓ of the way around, so the eclipse that we had seen in Africa on June 30, 1973, was repeated for us in the Western hemisphere on July 11, 1991. The name saros wasn't given until the seventeenth century, when Edmond Halley supplied it. Halley referred to a misunderstood phrase in writing by the ancient-Roman scientist Pliny in choosing the word.

It must have been difficult for the ancients to find out about this saros interval, because many eclipses were on the wrong side of the world for them and because others were lost to clouds. Yet careful record keeping was obviously sufficient. It has even been suggested that one use of the giant set of stones at Stonehenge in England was as an eclipse predictor. Since the stones showed how the positions of the sun and the moon changed over time, the times when the sun and moon would be in a line could also be predicted. Though the idea is still controversial, markers could have been moved in a set of holes still known at Stonehenge to keep track of lengthy eclipse periods.

Myths from the Far East also indicate that the repetition of eclipses was understood. A traditional story is that the Chinese court astronomers Hsi and Ho overimbibed and failed to predict an eclipse, for which they lost their heads. This eclipse would have occurred at about 2000 BC, about when Stonehenge was being constructed, but the truth of the story is lost in the depths of time, and in the burning of all historical records that occurred in China. Often, the eclipse that calculations show took place on October 22, 2134 BC, is identified with the event, though the actual date is uncertain to within hundreds of years. Actually, Hsi-Ho was the name assigned to a sort of sun deity, with responsibility for preventing eclipses. They were supposedly from the Xsia dynasty, which is not historically established. In the words of F. Richard Stephenson of the University of Durham, it is "a tale without any foundation whatever."

The earliest prediction of the actual path of an eclipse across the face of the earth was made by the remarkable astronomer Sir Edmond Halley for the eclipse of 1715. This first known prediction of the actual path of totality came 10 years after Halley used the law of gravity advanced at his behest by Sir Isaac Newton to discover the periodicity of the comet that now bears Halley's name. So Newton's law of gravity not only revealed the truth about an important comet but also allowed us to obtain the accuracy needed to predict eclipses. The 1715 eclipse crossed southern England, and reports from various sites around the country are still being examined to find out just where the edges of the path were. Even data almost three hundred years old should be accurate enough to tell us whether an eclipse was total or not! Determining the path can tell us the size of the sun, which is significant because even small changes of the sun's size could lead to changes in our climate here on earth. Some scientists have suggested that small changes in the sun's size have been measured from eclipse to eclipse, but my own measurements at the total eclipse in Papua New Guinea in 1984 indicated that the uncertainties in measuring eclipse durations are too great to resolve the issue.

What was the earliest eclipse recorded? Some scientists think that an engraved tablet now in Damascus, Syria, records the eclipse of 1223 BC. If this translation is accurate, the tablet is the only surviving astronomical observation written in the Ugaritic language of Babylonia. The tablet refers to an event occurring at sunset, and it may not have been an eclipse after all. Since we don't know the Ugaritic names of the planets nor do we know the calendar they used, the identification of the tablet with an eclipse remains speculative. And even if it was an eclipse, it might not have been total. In 1993, an Orientalist and a historian of science from the University of Chicago concluded that the text does not correspond to a solar eclipse at all.

The ancient sightings of eclipses are still significant to us. To track down why, I went to see Stephenson in his office at the University of Durham. He has championed the use of ancient texts to determine the speed at which the earth rotates. "If the earth had rotated a little faster or slower than it does now," he explained, "the path of the eclipse would cross a different region of the earth. So we can find out very accurately how fast the earth had rotated on average ever since." When I was there, scientists were visiting him from Kuwait, Saudi Arabia, and other countries from which ancient chronicles sometimes reveal eclipse records. They continue to search for additional evidence of ancient eclipses in old chronicles. Said Said from the King Saud University in Riyadh explained, "This is a thorough search, volume by volume, of all sorts of chronicles that are well known by historians. The chronicles are in Cairo, Damascus, Baghdad and, for Islamic Spain, Córdoba. Nobody has done this before for Arab chronicles." The only way to find references, he explained, is to "wade through it." Stephenson disappointed me, though, when he told me that "Written accounts are common but it is very rare that there are illustrations." Also, the references aren't usually direct. "They sometimes express themselves in a poetic way." Also, eclipse illustrations were "virtually unknown in China too. It is the written word that seems much more potent to a chronicler."

Stephenson has investigated the details of the northern limit of the 1715 eclipse, which has been used to find the eclipse diameter. "I visited the northern site – in Darington – where the Hall is still standing. The easiest way in a small village is to approach the minister. He mentioned a couple there who took us on a tour to see the Hall where the marginal north limit was and also gave us a copy of the deed of sale of the Hall." Still, Stephenson concluded, "The observation is of interest but more of curiosity value than of real scientific interest. How do you interpret a marginal observation?" Stephenson, along with colleagues John Parkinson and

Leslie Morrison, found no trend in the solar diameter. "When you get down to things, there isn't in that material what you want."

The *Nuremberg Chronicle*, published in Nuremberg, Germany, five hundred years ago in 1493, has many references to eclipses and to comets. But the accuracy of the information declines the farther back in time it comes from, and it is impossible to determine whether the discussions of astronomical events are literal or figurative. Prof. R. J. M. Olson, an art historian from Wheaton College in Norton, Massachusetts, and I (JMP) travelled to Nuremberg to examine the handwritten manuscript from which the *Nuremberg Chronicle* was printed and from which the woodblock illustrations may have been copied. We found the sketches in these Exemplars too quickly drawn to be astronomically accurate. So the resulting woodblocks may be beautiful, but cannot be relied on for scientific accuracy.

Kevin Pang of Caltech's Jet Propulsion Laboratory has been examining ancient Chinese records, again to find eclipses that can reveal how fast the earth has been rotating. With Chou Hsiang, a UCLA professor of Chinese, he analyzed a sixth-century BC history book that contained old stories and dated one of the eclipses as that of October 16, 1876 BC. If this dating is correct, a day on earth was then 70-thousandths of a second shorter than today. Still, the written reference occurred more than a thousand years after the event. Pang, Chou, and colleagues have also found a reference on old tortoise-shell chips known as "dragon's bones" or "oracle bones" that read "Three flames ate the sun, and big stars were seen." They identified the event with the total eclipse of June 5, 1302 BC. The length of the day would then have been 47-thousandths of a second shorter than today.

Solar eclipses are so spectacular on earth because of a happy coincidence: the moon is both 400 times smaller than the sun and 400 times closer to the earth. If it were merely 400 times smaller, it would appear 400 times smaller in the sky. But because it is also the same 400 times closer, the moon takes up in the sky almost exactly the same angle as the angle the sun takes up. Thus when the moon passes in front of the sun, it blocks out the sun almost exactly. For no other solar-system planet and moon does such a lucky coincidence exist.

Largely because the moon's orbit around the earth is elliptical instead of round, the moon is sometimes farther away from the earth than average and sometimes closer. Thus it is sometimes smaller than average and sometimes larger. When it is larger than average and goes in front of the sun, the solar eclipse that results is relatively long. When the moon is smaller than average and goes in front of the sun, a ring of sunlight remains around the silhouette of the moon. Those eclipses are called "annular," because of this annulus (ring) of sunlight.

Solar eclipses always occur at new moon, since the lunar phase we call new moon simply results from the moon's being so close to the sun in the sky that only the moon's far side is receiving sunlight. Most months, however, the moon is above or below the sun's direction in the sky, so we do not have an eclipse. But no fewer than two nor more often than five times a year, the moon partly covers the sun, giving us a partial eclipse. And on the average of once every 1½ years, the earth, moon, and sun are in such a precise line that the eclipse is total or annular.

Only the total eclipses are spectacular, because the sun is so much brighter than the moon, the stars, or the everyday blue sky that shutting off only part of its light may not be obviously noticeable. The sun is a million times brighter than the full moon, for example. If even 1% of the everyday solar surface remains visible, the earth still receives 10,000 times more light than it does during full moon, and the sky remains blue and bright. Only when the sun is entirely covered – at a total solar eclipse – does the sky go dark. And only in such a dark sky can we see the faint outer parts of the sun. The halo of light that surrounds the sun, the corona, then appears readily to our view. It is about the same brightness as is the full moon, and is equally safe to look at. By contrast, during a partial or annular eclipse, the remaining part of the solar surface is too bright to look at safely without taking special precautions.

The region on the earth from which the sun is blocked at a total eclipse is in the shadow of the moon. The moon's shadow is long and tapering, and only barely reaches the earth. For the July 11, 1991, total solar eclipse, the shadow was never more than 160 miles (258 kilometers) across. As the earth rotated and as the shadow moved through space, the shadow swept out a region thousands of miles (kilometers) long. It began in the ocean west of Hawaii, crossed the Big Island of Hawaii and more Pacific Ocean, reached Mexico, and continued through Central America and South America. For thousands of miles (kilometers), to either side, people saw the sun partially covered by the moon, but these partial phases are not spectacular. Thus the coterie of people seeing a total solar eclipse is relatively small.

A lunar eclipse, on the other hand, is visible to people over about half the surface of the earth. They merely have to be able to see the moon at a time when the earth's shadow falls on it. Though it is by no means spectacular, a total lunar eclipse can be beautiful and awe-inspiring as the moon darkens until it is dimly seen as a faint reddish object in the sky. It gets its red color because the small fraction of the sun's light that passes through the earth's atmosphere and strikes the moon is reddish. The earth's atmosphere scatters the blue light better than the red, making the sky blue for other people and leaving only the red to get through. Sunsets are red for the same reason.

It is easy to understand why total eclipses are so memorable in ancient records. The darkening of daytime until the solar corona, the stars, and the planets are visible is so dramatic that even today's observers often regain a tremendous sense of awe. Mark Twain put the idea to memorable use in his novel *A Connecticut Yankee in King Arthur's Court*. His hero supposedly fell asleep and awakened in King Arthur's ancient England. "But all of a sudden I stumbled on the very thing, just by luck. I knew that the only total eclipse of the sun in the first half of the sixth century occurred on the 21st of June, AD 528, OS, and began at 3 minutes after 12 noon. I also knew that no total eclipse of the sun was due in what to *me* was the present year – *i.e.*, 1879. So, if I could keep my anxiety and curiosity from eating the heart out of me for forty-eight hours, I should then find out for certain whether this boy was telling me the truth or not." Fortunately, to save himself, our hero remembered, "It came into my mind, in the nick of time, how Columbus, or Cortez, or one of those people, played an eclipse as a saving trump once, on some savages, and I saw my chance." "Go back and tell the king," our hero said, "that at that hour I will smother the whole world in the dead blackness of midnight; I will blot out the sun, and he shall never shine again; the fruits of the earth shall rot for lack of light and warmth, and the peoples of the earth shall famish and die, to the last man!" Finally, the king gave in and spared our hero, who had to stall until the eclipse was total. "It got to be pitch-dark, at last, and the multitude groaned with horror to feel the cold uncanny night breezes fan through the place and see the stars come out and twinkle in the sky. At last the eclipse was total, and I was very glad of it." Finally, "I said, with the most awful solemnity: 'Let the enchantment dissolve and pass harmless away!' When the silver rim of the sun pushed itself out, a moment or two later, the assemblage broke loose with a vast shout and came pouring down like a deluge to smother me with blessings and gratitude . . . ."

Twain's description of an eclipse is accurate, but eclipses in stories are sometimes not true to life. In H. R. Haggard's famous 1885 story *King Solomon's Mines,* an eclipse is similarly used to gain influence. "Tell them that you will darken the sun to-morrow." I am sympathetic to their worry that "suppose the almanac is wrong." The reply that "Eclipses always come up to time" was surely reassuring. However, the author obviously had never seen a total eclipse, because he allows an hour of darkness for people to sneak away.

Mark Twain's and Haggard's fictions are fun, but a real story was similar. Christopher Columbus, nearly 500 years ago, had actually used a total eclipse of the moon to his own advantage when he was stranded in Jamaica. He knew from the tables of astronomical positions he carried that a lunar eclipse would take place on February 29, 1504. He scheduled a meeting with the local chiefs to take place during the eclipse. He used the eclipse to bargain for a restoration of food deliveries.

The Shawnee Prophet known as Tenskwatawa, who was an important leader of Native Americans in Ohio and Indiana in the early 1800s, used his knowledge of a forthcoming eclipse to solidify his power. When challenged by William Henry Harrison, then governor of the Indiana Territory and later to be President of the United States, to "cause the sun to stand still" or some other miracle, Tenskwatawa offered to "blacken the face of the sun." And so he did, or seemed to, though only for the duration of the July 16, 1806, total eclipse.

Farther back in time, people were often surprised by eclipses. The Greek historian Herodotus in 430 BC described how, on May 28, 585 BC, a total eclipse occurred during a lengthy war between the Lydians and the Medes. The Lydians and the Medes quickly reached a peace agreement. Herodotus also wrote that the Greek astronomer Thales had predicted the year of the eclipse. Still, the armies probably did not know what was coming. In any case, as we see from today's supermarket astrology predictions, people are often superstitious even when scientific truth is known.

If we count partial, annular, and total eclipses, as well as all types of lunar eclipses, there can actually be as many as seven eclipses in a given year. But total solar eclipses occur only about every 18 months. If we were to stay in one position on the earth, a total solar eclipse would come to us only about every 330 years. Now that we are a mobile society, it is much easier to travel to eclipses. Tens of thousands of tourists now make it their practice to assemble at places where eclipses will pass. Though often called "eclipse chasers," they are actually the opposite, for eclipses go too fast to chase. The moon's shadow whizzes across the earth's surface at speeds greater than 1,000 miles per hour.

By keeping track of solar and lunar eclipses, ancient peoples were able to figure out intervals by which eclipses repeat. Whether or not the impressive stone pylons erected over 4,000 years ago at Stonehenge, in south England, are used as part of a giant calendar to predict eclipses is controversial. There is little disagreement, however, that the stones of Stonehenge were used to mark the positions in the sky where the sun and the moon reached their extreme locations to the north or south. Thirty years ago the astronomers Gerald Hawkins and Fred Hoyle suggested ways in which a set of holes outside the main ring could be used to mark eclipse cycles. Moving stones from hole to hole at fixed intervals could have kept track of eclipses. Of course, the people at Stonehenge would have seen mainly partial eclipses, and even many of those would have been lost to cloudy weather. Since we have no written records from the time, it is up to each of us to decide whether the Stonehenge erectors really predicted eclipses or whether we are merely smart enough to think of ways that they could have done so.

Much closer to our time, it is clear that the Maya in Mesoamerica knew how to keep track of eclipse cycles and how to predict eclipses. I asked my colleague Sam Edgerton, of the Clark Art Institute in Williamstown, Massachusetts, to tell me about the Mayan records. Sam now often forsakes his former primary field, the study of art in Renaissance Italy and the discoveries made then by Galileo and his contemporaries, for investigations in the jungles of Central America. He explained to me that the Dresden Codex is a book of hieroglyphics giving dates from about AD 700 until it was written in about 1400, one of only four that had not been burnt by the conquistadors. The Codex got its name from its current location in Dresden, Germany, where it was noticed in the eighteenth century. It may have been brought to Europe in 1519 by Cortés as a gift to the emperor Charles V after the conquistadors ravaged the Mayan civilization. "In order for them to go from the ground to the stars in this agricultural community, all the metaphors, all the iconography of existence, of life and death, were based on the metaphors of agriculture. They had an idea of things cycling back. The word Maya itself means "people of the cycle."

"The Maya developed a very complex religion, thought, mythology, and a language system that was to communicate this, to keep a record of it. The Maya were concerned that events that happened in historical time repeated themselves. And they should be ready for these, because many of the events were disasters: the coming of floods, the coming of dry seasons, the coming of one terrible consequence or another." Sam went on to explain, "So the Maya took their relationship to the stars and made it a little more scientific in that they organized it around a record-keeping system. Their writing of glyphs and their inventing of an arithmetical system were built in order to keep track of the historical record that they imagined began at a mythological date way back at a time that they fixed. And everything progressed on a date calendar afterward.

"So, getting to why the Maya would keep track of eclipses, it was only one of a number of astronomical events that interested them: the rising and setting of Venus, for example, was very important. Venus was seen as the herald of the sun. As Venus came up and led the sun up, Venus was seen as a dog, literally, leading his master up and taking his master down into the underworld.

"The Dresden Codex was one of a number on bark paper. They folded it like an accordion. This one was about 20 feet long. They would keep track, for example, of the vernal equinox, which was very important because then you would tell the farmers to start planting.

"The Dresden Codex is, in fact, an almanac used to predict the rise and fall of Venus and of eclipses. But," he explained, "it's not an eclipse record as such, but a record of times when eclipses *can* happen."

Sam showed me how the glyphs give the number of days in the intervals between eclipses. "In a large Mayan number like this, the bottom row has 3 bars, which is 15, and the two dots make 17. The bar on the top with three dots makes 8. In the top row, you multiply by twenty, just as you know that in the decimal system that 1. is 10 times greater than .1; twenty times 8 is 160, plus 17 makes 177. Sometimes the top shows 7, and twenty times 7 is 140, plus 18 in the bottom row makes 148." We now know that these intervals of 177 days and 148 days occur because they correspond to when the sun and moon are at the positions in their orbits where the orbits apparently cross as seen from earth. Indeed, these intervals also describe eclipses of our own times. We say that they link "eclipse seasons." The next total eclipse after July 11, 1991, took place about twice 177 days later, on June 30, 1992. There will be an interval of 177 days between the May 10, 1994, annular eclipse and the November 3, 1994, total solar eclipse.

But the Mayan eclipse dates give merely the times when eclipses were possible, not when they would actually occur. So they would tell the people when to be ready for eclipses, but were not actual predictions. Still, if the Arawak Indians who met Columbus in Jamaica had been Maya, they might have known almost as much about predicting eclipses as Columbus did.

Sam also explained the glyphs to the left of the page. "Below the sky bar in the middle, we see dark and light bands. The fact that you have the dark band and the light band shows that you are confused by the sun and the moon. Now, right below that, you see the same sky bar with the Venus symbol on the left. The two black and white things with the crossed bones. Notice that underneath that is the serpent with an open mouth. That is the image of the earth, the earth serpent, like a celestial monster. From the sky bar, down descends the black/white, meaning the sun and the moon, and it's about to be swallowed up by the earth's serpent."

Within every 18-year interval, we have a series of solar eclipses, some long and some short. The longest eclipse ever possible has 7 minutes 31 seconds of totality. The June 30, 1973, eclipse that crossed Africa was the only eclipse this century to exceed 7 minutes. One saros later, after waiting 18 years 11⅓ days, we had the long eclipse of July 11, 1991, whose longest totality was just under 7 minutes. This long eclipse in the saros won't repeat until July 22, 2009.

The last few years have been very exciting for studies of the sun, because the amount of solar activity was at a near-record high. The easiest way to see the solar activity is by observing sunspots, which wax and wane in number with a period of about 11 years. The peak of this sunspot cycle, the second highest ever recorded since sunspots were discovered by Galileo in 1610, occurred in July 1989. But solar activity continues to be high for a couple of years after the sunspot peak, and powerful flares more often occur in this post-maximum period. The spring of 1991 saw the earth bombarded with particles and radiation from a series of powerful flares on the sun. They caused the aurorae to be seen more widely than usual, and threatened power lines with surges of current. The sunspot number declined rapidly in 1992, as we approach the 1995 sunspot minimum.

The corona is but one of the solar phenomena that changes with the solar-activity cycle. At solar minimum, when the number of sunspots we see is close to zero, the corona extends mainly outward from the sun's equator. The sun's magnetic field binds the coronal gas into beautiful shapes called streamers. Some are wide at their base and curve down into points; they resemble ancient helmets, and so are called "helmet streamers." At the sun's poles, thinner streams of gas can be seen following the sun's magnetic field, like iron filings following the magnetic field of a bar magnet.

As solar activity increases, the number of streamers increases and they appear at a wider range of solar latitudes. When we see them in projection, as three-dimensional spikes projected against the sky, the streamers then make the solar corona appear round. Thus the round corona at the 1980 eclipse contrasted with the irregularly shaped corona at the 1984 eclipse. The presence of the solar-activity maximum made the 1991 eclipse especially interesting. Often, there were a hundred or more sunspots on the face of the sun, and the corona was very round. We will be back near the peak of the sunspot cycle for the 1999 eclipse, which crosses Europe.

Further, the decade of the 1980s had been one of tremendous technological advance. Instead of film, for which only about one out of every 100 photons of light that hits contributes to the image, astronomers mainly now use electronic imagers that sense about half the photons. Thus our cameras – using these new "charge-coupled devices," or CCDs – are about 50 times more sensitive. In one sense, in four minutes of total eclipse, we can do 4 times 50, or 200 times, the observations we could do ten years ago. It is like having a 200 minute, or 3 hour 20 minute, eclipse. We (JMP and students and colleagues) could thus plan to make a map of the temperature of the corona whereas ten years ago, with the same method, we could have merely determined the temperature at one or two points. A grant from the Committee on Research and Exploration of the National Geographic Society aided us in making our expedition.

Another major advance in astronomy in the past decade was the development of devices that could sense the infrared part of the spectrum far better than before. These infrared devices could not map as finely as CCDs, but they were still far better than observing point by point. The 1991 eclipse, astonishingly, linked the development of these new infrared imagers with the ability to observe the infrared, which does not come down to ground level. The eclipse, fortunately, passed directly over the top of Mauna Kea, where more of the world's largest telescopes exist than at any other location. Only every few decades does a total eclipse pass over a modern major observatory. For some of the infrared-imaging experiments, it wasn't even necessary to use the big telescopes, since the sun is so bright compared with the stars and nebulae usually observed. We discuss this magnificent eclipse in Chapter 8.

Advances in theory also called for making maximum use of the long total eclipse. Since observations from a spacecraft showed 15 years ago that our old ideas were wrong about how the solar corona was heated, new models have been developed. We now think that the sun's magnetic field is important in channeling special kinds of waves from under the sun's everyday surface into the corona. Only with such waves carrying energy upward can we account for the rise in temperature from the surface's 6,000°C to the corona's 1,000,000°C. The other major experiment my students and I tried to carry out in Hawaii was meant to observe the effects of one such wave of this magnetic type. It was predicted that the waves would move along the loops of gas that make up the corona and make them shake and vary slightly in brightness with a period of about 1 second. We had special apparatus to record the brightness 30 times a second, so we could study whether the loops were varying rapidly.

Another model for how the corona gets so hot is that a lot of tiny solar flares are going off very often. One international set of astronomers from the US, Canada, and France used the giant Canada–France–Hawaii telescope on Mauna Kea to make high-speed images with very excellent spatial resolution. Studies of the details and how they change should reveal whether these flares occur. A set of similar, smaller telescopes spread out over the eclipse path from Hawaii to Mexico to Brazil was to show larger-scale changes in the few hours that the eclipse took to travel across the earth.

Most of the observations made at the eclipse could not be carried out in any other way. We have not yet developed spacecraft that can study the lower and middle corona as well as it can be studied at eclipses. Even the Solar Maximum Mission that was aloft from 1980 to 1989 had to mask out not only the sun's ordinary surface but also its lower and middle corona in order to detect the outer corona. And we didn't even have a solar satellite aloft during the eclipse. Today, the Japanese Yohkoh ("Sunbeam") satellite is aloft, making x-ray (but not visible) images of the sun, which show the corona as well as explosions known as solar flares. We can use eclipse images to compare with the satellite images taken in other parts of the spectrum. Eclipse studies will still be useful for decades.

In any case, eclipses are spectacularly interesting and beautiful. To aid you in watching and in photographing them, we present this book.

CHAPTER ONE

ECLIPSES OF THE MOON

How lunar eclipses happen

A lunar eclipse is visible when the moon passes into the earth's shadow. Because the moon is actually in darkness, you can see a lunar eclipse from anywhere that the moon is in the sky at the time. Consequently, over the years at any given location, you can see far more eclipses of the moon than of the sun.

Figure 1.1 shows that the shadow of the earth, like any shadow from a light source of appreciable size, consists of two parts: an inner, uniformly dark part called the *umbra,* and an outer, fuzzy, gray part called the *penumbra* that gets gradually lighter toward the edges. (The same is true of the shadow that your hand casts in sunlight or the light of a single light bulb; try it.) Figure 1.2 shows the moon's passage through the earth's shadow as seen from earth; the moon enters first the penumbra, which causes it to dim somewhat, and then the umbra, which plunges it into almost complete darkness. If the only light reaching the moon were direct sunlight, the moon would be completely black and invisible while in the umbra, but in fact a certain amount of light gets bent around the earth, by the atmosphere, and bathes the moon in a dim, coppery glow. The brightness of this glow is unpredictable, since it depends on the distribution of clouds and other obstructions (such as volcanic dust) in the earth's atmosphere.

Dates, times, and visibility

Table 1.1 lists the lunar eclipses that will occur during the remainder of the twentieth century. Nearly two-thirds of these are *total* eclipses; that is, at some point during the eclipse the moon is completely within the umbra.

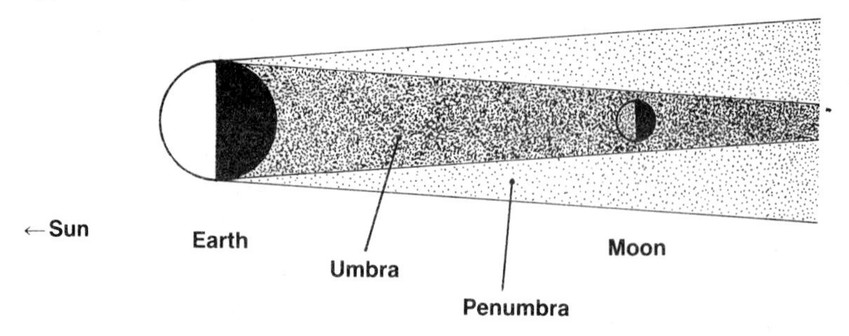

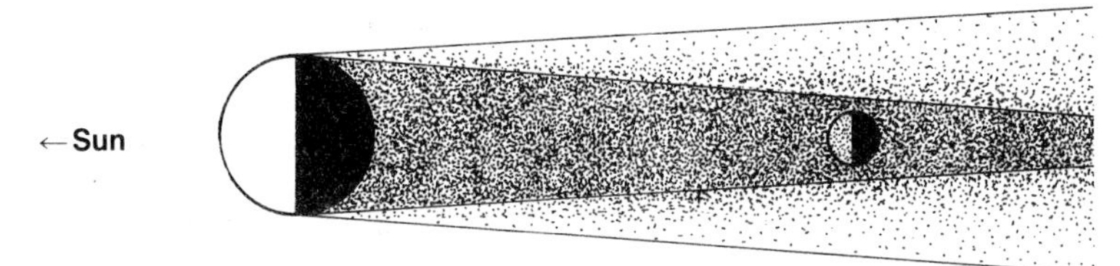

Fig. 1.1. A lunar eclipse.

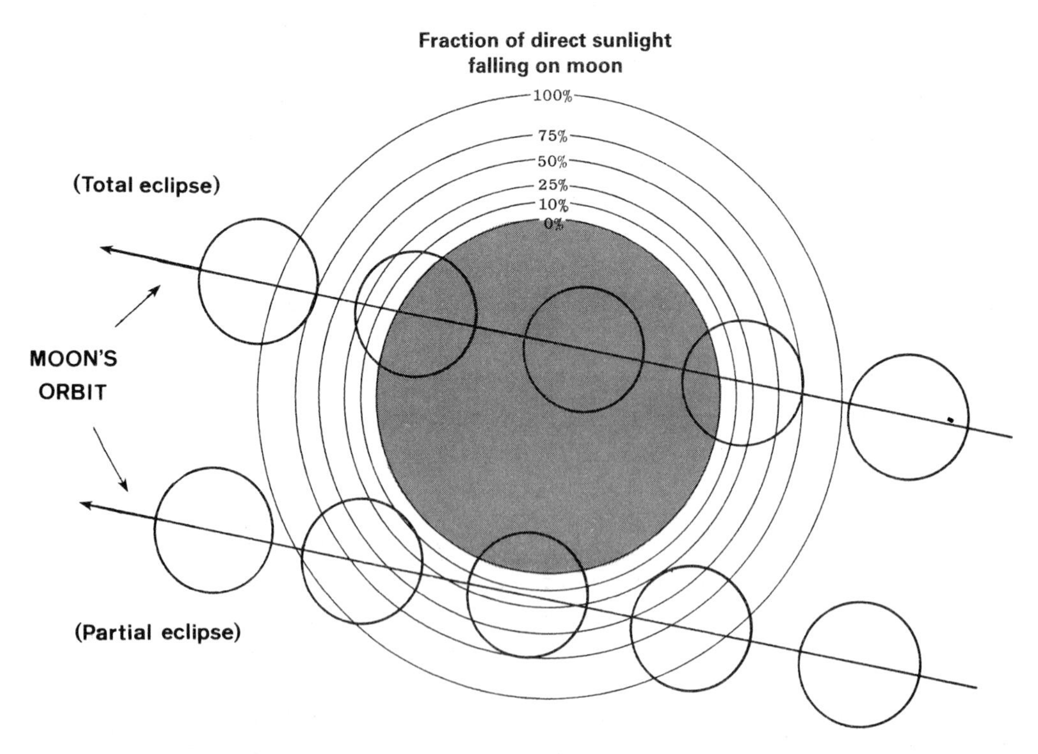

Fig. 1.2. Light intensity on the moon during a lunar eclipse, as seen from the earth. (The shaded area is the umbra.)

Table 1.1. *Umbral lunar eclipses, 1993–2000*

All times are in Universal Time (UT). Eastern Standard Time is 5 hours earlier; Eastern Daylight Time is 4 hours earlier. Subtract an additional hour for Central Time, 2 additional hours for Mountain Time, and 3 additional hours for Pacific Time.

Date	Partial or total	First contact	Mid-eclipse	Last contact
June 4, 1993	Total	11.12	13.01	14.50
Nov 29, 1993	Total	04.40	06.25	08.10
May 25, 1994	*Partial*	*02.39*	*03.31*	*04.23*
April 15, 1995	*Partial*	*11.42*	*12.18*	*12.54*
April 3–4, 1996	Total	22.22	00.10	01.58
Sept 27, 1996	Total	01.13	02.54	04.35
March 24, 1997	*Partial*	*02.59*	*04.40*	*06.21*
Sept 16, 1997	Total	17.08	18.46	20.24
July 28, 1999	*Partial*	*10.22*	*11.33*	*12.44*
Jan 21, 2000	Total	03.03	04.44	06.25
July 16, 2000	Total	11.58	13.56	15.54

The rest are *partial*, in which the moon grazes the umbra and is never more than partly immersed in it. There will also be frequent *penumbral* eclipses, in which the moon passes through some part of the penumbra without touching the umbra; these are not listed in the chart because the dimming is so slight as to be almost imperceptible.

To find out whether you can see a particular eclipse from your location, you need to know the time at which the eclipse takes place and whether the moon will be in the sky at your location at that time. The times in Table 1.1 are expressed in Universal Time, UT. This time scale is close to Greenwich Mean Time (GMT), which is no longer used. Actually, astronomers now use a variation of Universal Time known as Coordinated Universal Time, UTC. But the time scales differ only by seconds, so we won't worry about such details.

To convert UT to your local zone time, you will have to add or subtract a fixed number of hours; Table 1.2 lists the conversions for most of the English-speaking world. If your time zone is not listed or you want to double-check your conversion, all you need is a shortwave radio, since international broadcasters use UTC when they give the time of day. In addition, very accurate UTC time signals can be obtained by telephoning 303 499-7111 in the USA.

To establish whether the moon will be in the sky at the time of the eclipse, all you need are the local times of sunrise and sunset. Since the moon is to be in the earth's shadow, the moon and sun will be in almost precisely

Table 1.2. *Converting Universal Time (UT)*

Add or subtract the requisite number of hours, as indicated. If the answer comes out negative, add 24 and count it as the previous day. If the answer comes out greater than 24, subtract 24 and count it as the following day.

Time zone	Local zone time =	
	Winter (standard time)	Summer (daylight saving time)
Great Britain	UT	UT +1
Continental Europe	UT +1	UT +2
South Africa	UT +2	UT +3
Australia (west coast)	UT +8	UT +9
Australia (center)	UT +9½	UT +10½
Australia (east coast)	UT +10	UT +11
New Zealand	UT +12	UT +13
North America:		
Atlantic Time Zone	UT −4	UT −3
Eastern Time Zone	UT −5	UT −4
Central Time Zone	UT −6	UT −5
Mountain Time Zone	UT −7	UT −6
Pacific Time Zone	UT −8	UT −7
Alaska Time Zone	UT −9	UT −8
Hawaii Time Zone	UT −10	No d.s.t.

Remember that the summer in the northern hemisphere is, roughly, June through September, while summer in the southern hemisphere is, roughly, December through March.

opposite directions from the earth – which means that when the sun is in the sky, the moon isn't, and vice versa.

Let's put all these ideas together with a concrete example. Suppose you want to observe the lunar eclipse of May 25, 1994, and you live in California. Daylight saving time will be in effect in May, so your conversion formula will be:

local zone time = UT − 7 hours

The eclipse lasts from 2.39 to 4.23 GMT, and using the formula we find that this is equivalent to 19.39 to 21.23 Pacific Daylight Time the previous evening (that is, 7.39 p.m. to 9.23 p.m. on the 24th). Since lunar eclipses occur at full moon, and the full moon rises as the sun sets, the eclipsed moon should thus be up in the sky to see. In New York (where local time = UT −4) the eclipse will last from 10.39 p.m. to 12.23 a.m. Observers in

Fig. 1.3. An example of what can be done even under adverse conditions and with relatively unsuitable equipment. This is the partial lunar eclipse of July 17, 1981, photographed by Michael Covington with a 100-mm telephoto lens at f/2.8 together with a ×3 teleconverter, exposing 1 second on Fujichrome 400, through a momentary break in the clouds.

Fig. 1.4. The lunar eclipse of May 24, 1975, as photographed by Douglas Downing at the prime focus of a 15-cm (6-inch) f/8 Newtonian. Three seconds on Fujichrome 100, 15 minutes after the end of totality. A small amount of the reddish-brown umbra is visible, along with the greatly overexposed penumbra.

England will see the eclipse begin at 3.39 a.m. local time, but the moon will have set before it ends at 5.23. (Remember that the full moon sets approximately at sunrise, or use the charts on pages 21–27 to determine whether the whole eclipse will be visible at your location.)

Photographic exposures

Once you've chosen an eclipse to observe, you'll have no trouble photographing it. The equipment needed to photograph lunar eclipses is the same as for photographing the moon generally; if anything, the requirements for sharpness are less stringent, since surface detail is of less interest (Figs. 1.3–1.6). You'll almost certainly want to photograph in color, in case the umbra turns out to be a striking copper hue (see Plate 14).

The main problem is that of determining exposure. During the partial phases of the eclipse, two approaches are possible: you can expose for the bright area outside the umbra, making the umbra itself look pitch-black, or you can try to get both the bright and dark parts of the moon into the film's sensitivity range. Appendix B gives exposure tables for both of these approaches; the second is considerably the more difficult of the two, since the umbra and penumbra differ in brightness by a factor of about ten thousand. The dimmest part of the penumbra is the area closest to the umbra, so the best conditions for success occur when the moon is almost completely inside the umbra and only a small sliver of penumbra remains. Naturally, it helps if the film has good exposure latitude.

Correct exposures for the totally eclipsed moon are unpredictable because the amount of light reaching the moon through the earth's atmosphere cannot be predicted in advance; some lunar eclipses are relatively bright, with the moon continuing to glow light orange even when eclipsed, while at other times the eclipsed moon is so dark as to be invisible even in a telescope. Two tables in Appendix B give suggested exposures for relatively light and relatively dark eclipses, but these are only approximations; to be on the safe side, bracket your exposures.

One thing you'll notice from the tables is that exposures for photographing the totally eclipsed moon can be quite long – as much as several minutes at *f*/8, even on relatively fast film. This means that if you are using a relatively slow lens, you'll need to mount your camera on a telescope equipped with a clock drive to follow the earth's motion; otherwise, the limits in Table 3.3 (Chapter 3) apply (though they can be stretched a bit if critical sharpness is not necessary). With exposures of a couple of minutes or more, the difference between stellar and lunar drive rates can even become significant; the telescope will need to be driven about 3% slower than its usual star-tracking rate because the moon is moving continuously in its orbit.

Fig. 1.5. *The lunar eclipse of August 17, 1970 photographed with a 15-cm (6-inch) f/8 Newtonian telescope and 32-mm eyepiece coupled afocally to a rangefinder camera with a 45-mm lens, and focused by the hand-telescope method (p. 55). The exposure was 1/250 second, through high cirrostratus clouds, on Tri-X developed 7 minutes at 22°C (72°F) in D-19 1:1. (By Michael Covington, who was not quite 13 years old at the time)*

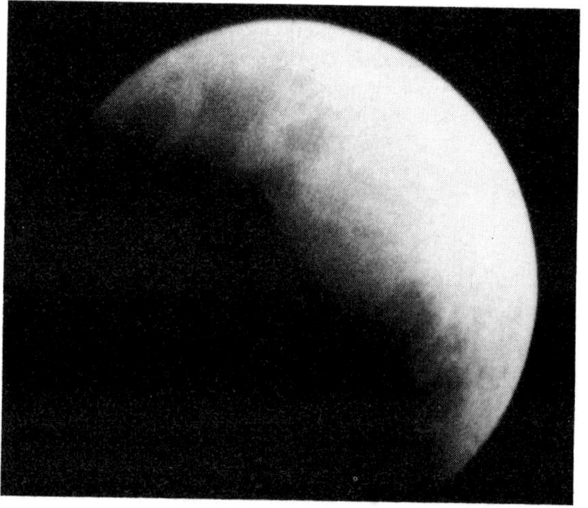

Fig. 1.6. *A partial phase at the total lunar eclipse of June 16, 1992. Jay M. Pasachoff used a 500-mm Nikon f/8 mirror telephoto lens on a Nikon F4 and bracketed two stops to either side of the spotmeter reading to make this photograph on Plus-X film.*

Fig. 1.7. The moon rising and coming out of eclipse on December 30, 1982. Akira Fujii made exposures at 5-minute intervals on Ektachrome 64 using a homebuilt 4 × 5-inch sheet film camera with a 400-mm lens at f/4.5, on a fixed tripod. The exposures varied from 1/2 second at f/5.6 for the thin crescent to 1/125 at f/11 for the nearly full moon. Note that the sun and moon move their own diameter across the sky in about 2 minutes. Thus Mr. Fujii took these images about every 5 minutes. The 18 images therefore took over one hour to produce. The field of view of the camera is just over 15° on the diagonal.

The dimness of the totally eclipsed moon does have one advantage; it makes it possible to take a picture of a star field with the moon in it, a feat that is normally impossible because the moon is too bright. Use a medium telephoto lens, mount the camera on a clock-driven telescope, and expose for two to five minutes; or put the camera on a fixed tripod and expose 10 to 30 seconds with the normal lens at *f*/2.

If your camera can make multiple exposures, you can take a quite striking picture of the progress of the eclipse by placing the camera on a fixed tripod and exposing once every five minutes or so as the moon moves through the field (Fig. 1.7 and Plate 13). Since the moon images will not overlap, each of them should be exposed normally. As an alternative, mount the camera on a clock-driven telescope, guide on a star (against which the earth's shadow is relatively stationary), and expose about once every 45 minutes; you'll get a photographic version of Fig. 1.2 or Plate 15.

Circumstances of the lunar eclipses†

In Figs. 1.8–1.14 each lunar eclipse has two associated diagrams. The top figure shows the path of the moon through earth's penumbral and umbral shadows. Above and to the left is the time of middle eclipse (MID), followed by the penumbral (PMAG) and umbral (UMAG) magnitudes of the eclipse. The penumbral and umbral magnitudes are the fraction of the moon's diameter immersed in the penumbral and umbral shadows respectively at middle eclipse. The geocentric distance between the moon's center and the shadow axis at middle eclipse (GAMMA) is in units of earth radii. To the upper right are the contact times of the eclipse, which are defined as follows:

P1 = Beginning of penumbral eclipse.
(Instant of first external tangency of moon with penumbra)
U1 = Beginning of umbral eclipse (or partial eclipse).
(Instant of first external tangency of moon with umbra)
U2 = Beginning of total umbral eclipse.
(Instant of first internal tangency of moon with umbra)
U3 = End of total umbral eclipse.
(Instant of last internal tangency of moon with umbra)
U4 = End of umbral eclipse (or partial eclipse).
(Instant of last external tangency of moon with umbra)
P4 = End of penumbral eclipse.
(Instant of last external tangency of moon with penumbra)

In the left middle is the angle subtended between the moon's center and the shadow axis at middle eclipse (AXIS), and the angular radii of the penumbral (F1) and umbral (F2) shadows. The moon's geocentric coordinates at maximum eclipse are given in the right middle. They consist of the right ascension (RA), declination (DEC), apparent semi-diameter (SD), and horizontal parallax (HP). The saros series for the eclipse is followed by a pair of numbers in parentheses. The first number identifies the sequence order of the eclipse in the saros, while the second is the total number of eclipses in the series. The Julian Date (JD) at greatest eclipse is given, followed by the extrapolated value of ΔT used in the calculations (ΔT is the difference between Terrestrial Dynamical Time and Universal Time, and

† This note and the maps are provided by Fred Espenak, NASA/Goddard Space Flight Center. On the maps, the white regions are the places from which the whole eclipse is visible.

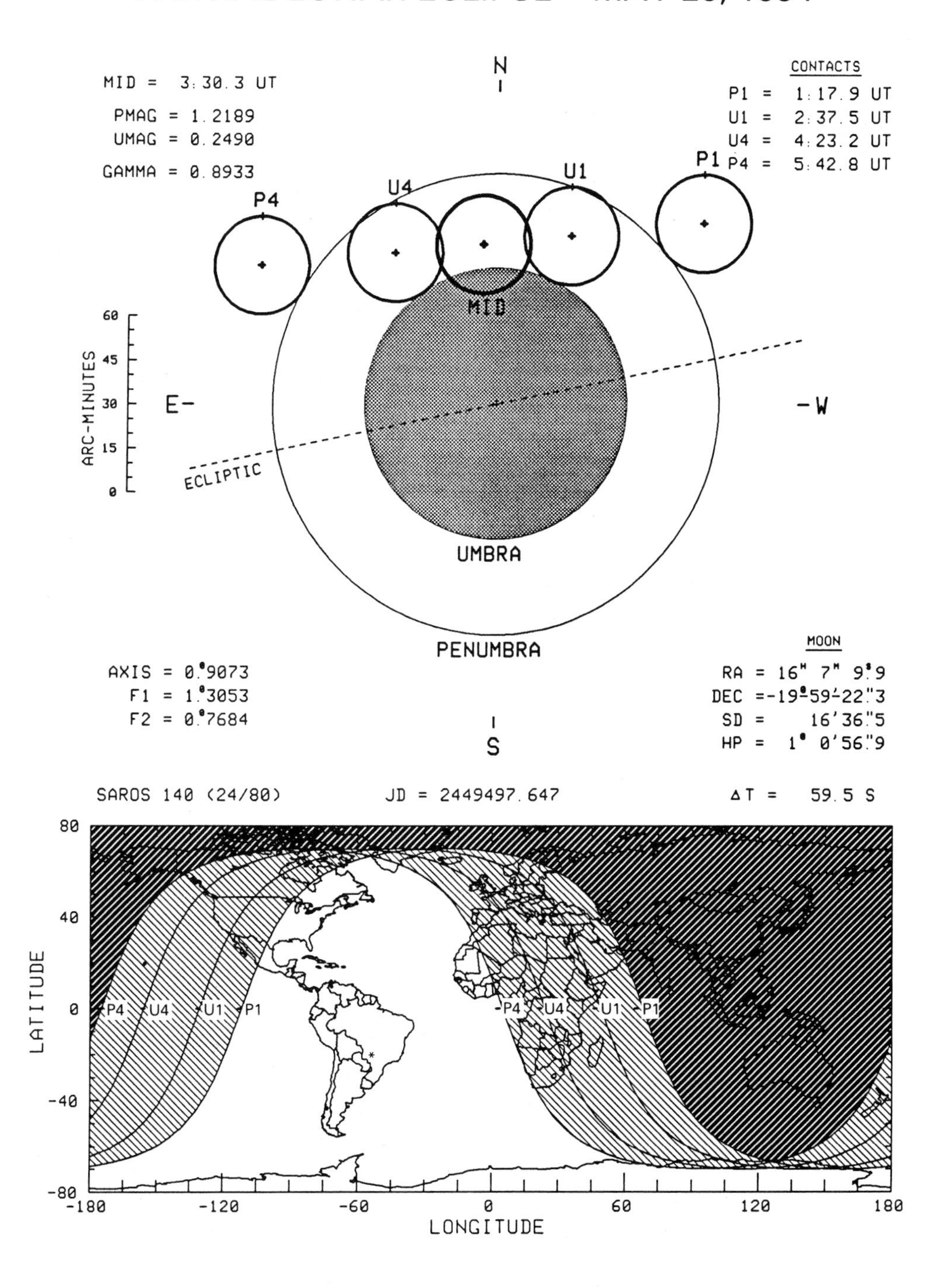

Fig. 1.8. *Partial lunar eclipse – May 25, 1994. Eclipse predictions and map courtesy of F. Espenak – NASA/GSFC.*

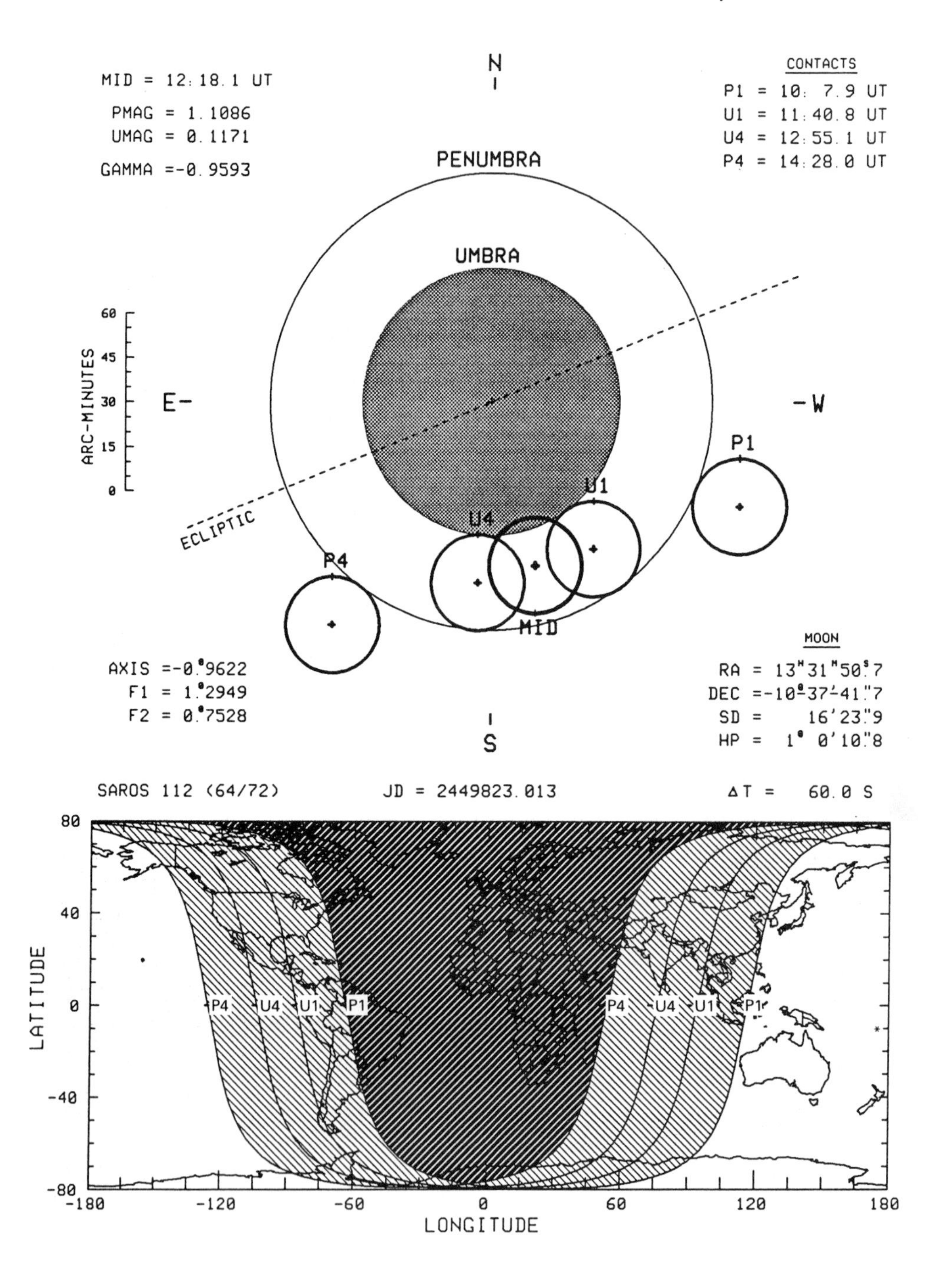

Fig. 1.9. Partial lunar eclipse – April 15, 1995. Eclipse predictions and map courtesy of F. Espenak – NASA/GSFC.

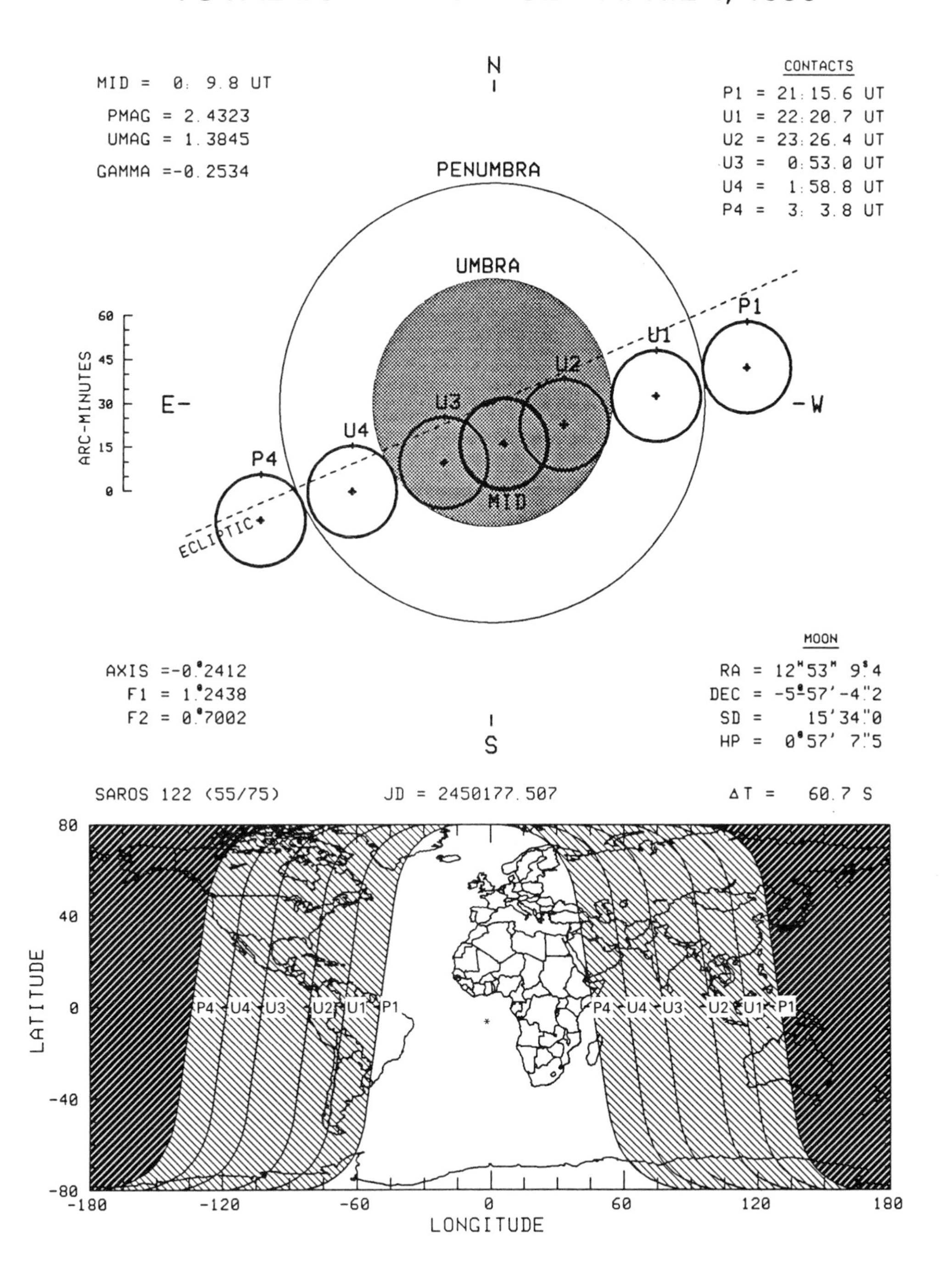

Fig. 1.10. *Total lunar eclipse – April 4, 1996. Eclipse predictions and map courtesy of F. Espenak – NASA/GSFC.*

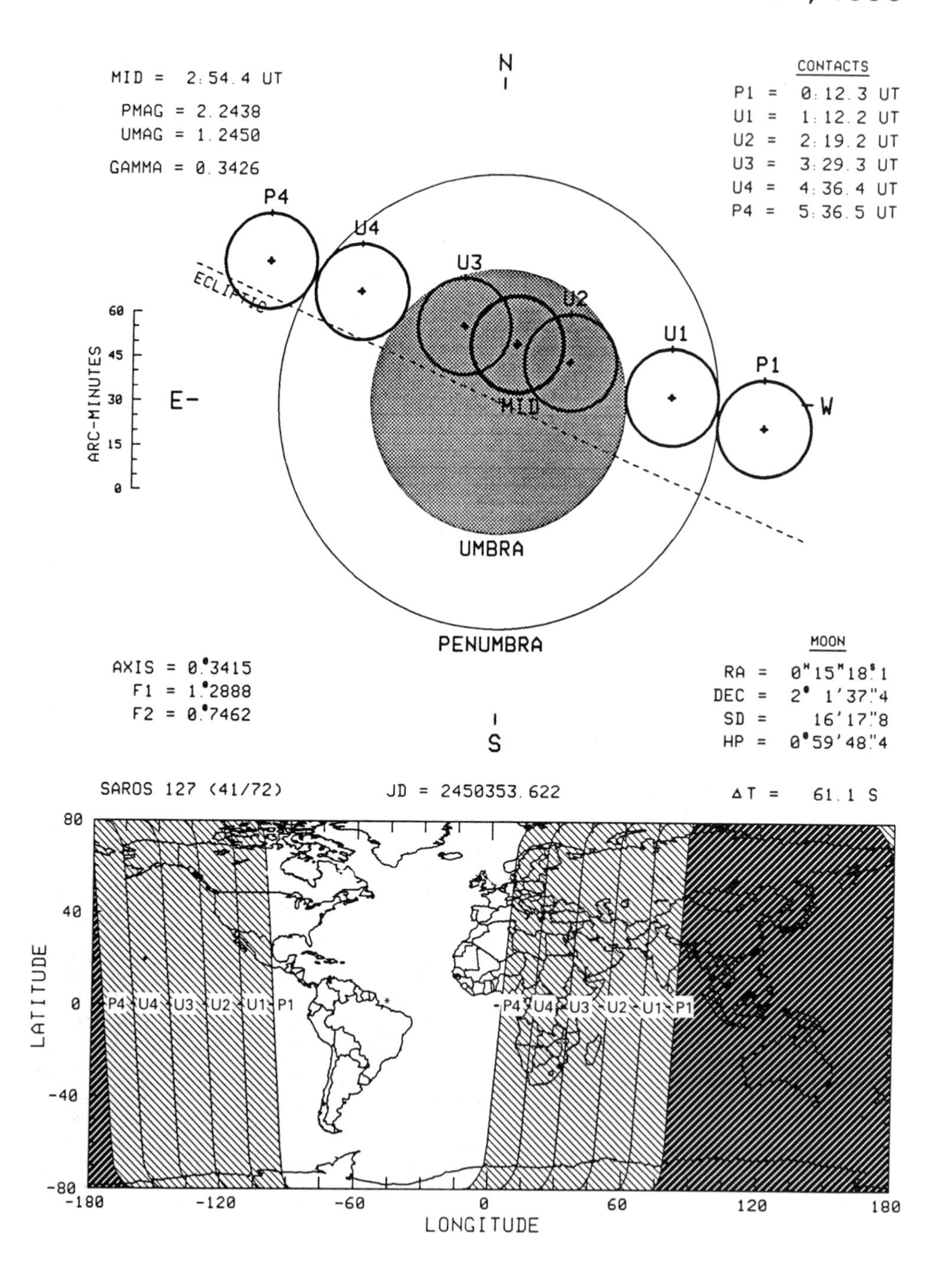

Fig. 1.11. Total lunar eclipse – September 27, 1996. Eclipse predictions and map courtesy of F. Espenak – NASA/GSFC.

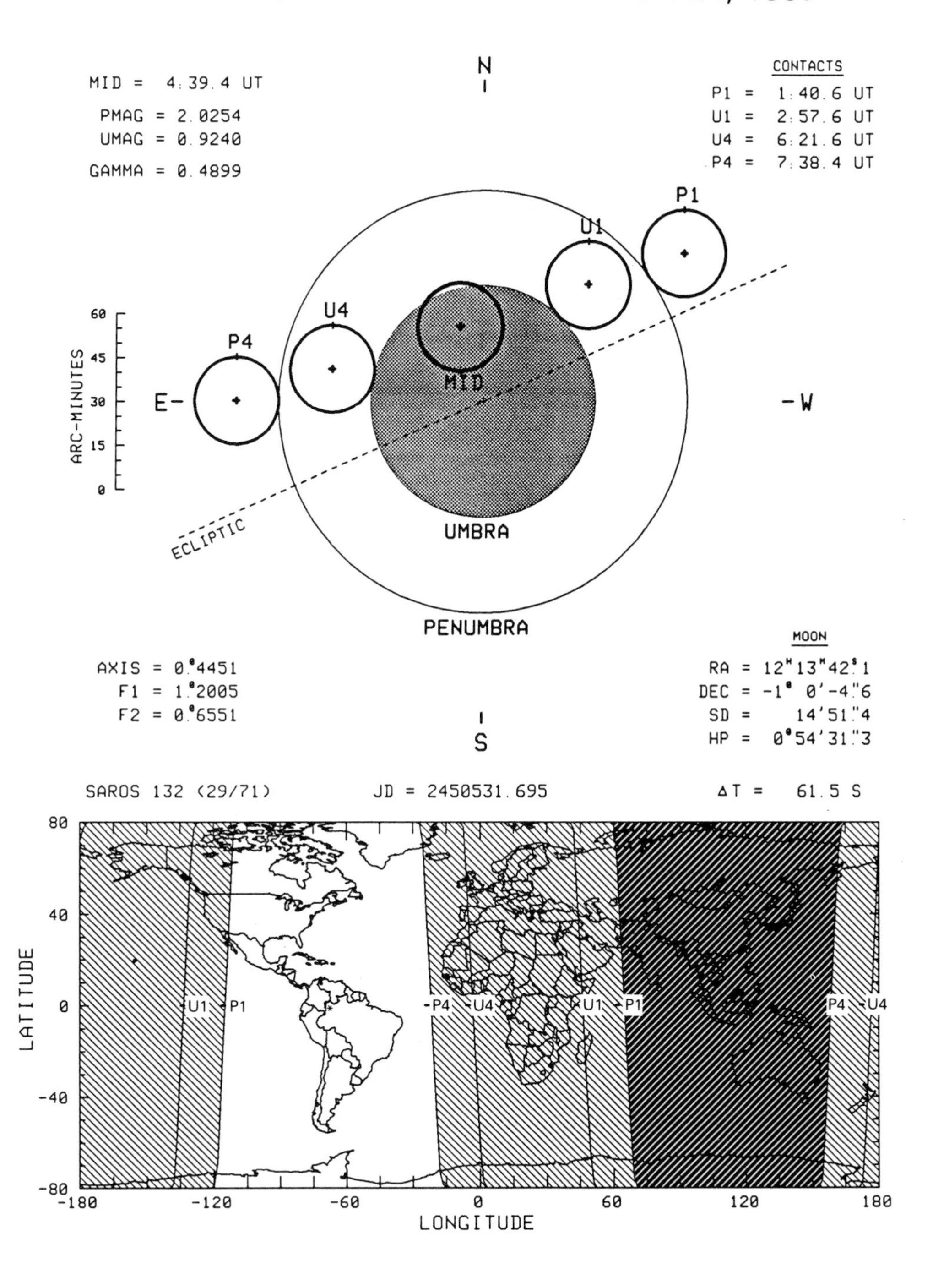

Fig. 1.12. Partial lunar eclipse – March 24, 1997. Eclipse predictions and map courtesy of F. Espenak – NASA/GSFC.

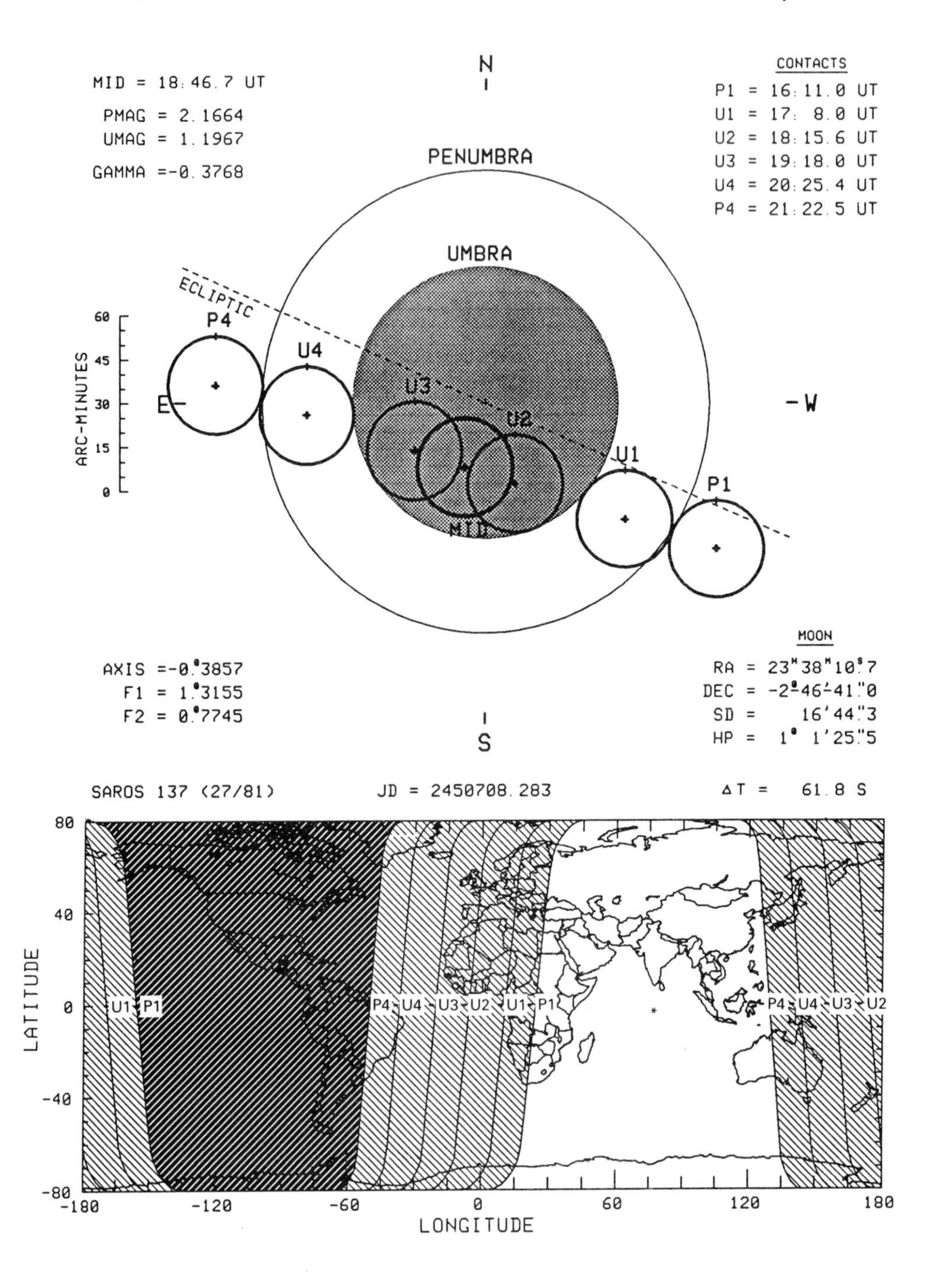

Fig. 1.13. *Total lunar eclipse – September 16, 1997. Eclipse predictions and map courtesy of F. Espenak – NASA/GSFC.*

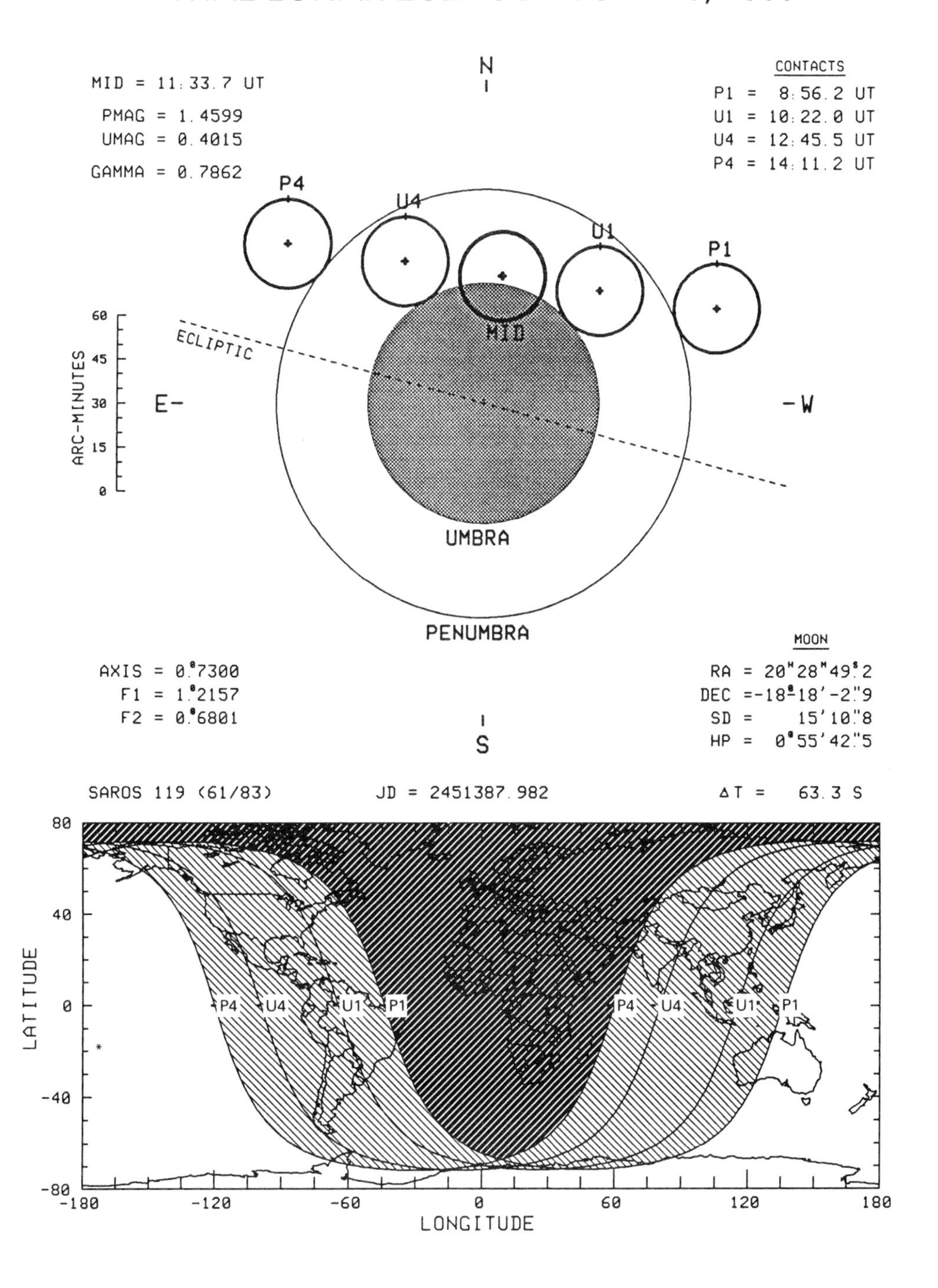

Fig. 1.14. Partial lunar eclipse – July 28, 1999. Eclipse predictions and map courtesy of F. Espenak – NASA/GSFC.

Julian Days are consecutively numbered days in a series whose origin is January 1, 4713 BC).

The bottom figure is a cylindrical equidistant projection map of earth which shows the regions of visibility for each stage of the eclipse. In particular, the moonrise/moonset terminator is plotted for each contact and is labeled accordingly. The point where the moon is in the zenith at middle eclipse is indicated by '*'. The region which is completely unshaded will observe the entire eclipse while the area marked by heavy diagonal lines will witness none of the event. The remaining lightly shaded areas will experience moonrise or moonset while the eclipse is in progress. The shaded zones east of the asterisk will witness moonset before the eclipse ends while the shaded zones west of the asterisk will witness moonrise after the eclipse has begun.

Chapter Two

Eclipses of the sun

Partial, total, or annular?

Solar eclipses result from the moon passing in front of the sun – that is, from the moon's shadow falling on the earth – and the appearance of the eclipse depends on the observer's location relative to the shadow, which is of course moving all the time. A total solar eclipse is seen as total only from a small region at the center of the shadow – the *zone of totality* – which is up to about 250 miles (400 km) wide (Fig. 2.1) and moves rapidly along a track several thousand miles (kilometers) long. Surrounding it is the *zone of partiality*, a couple of thousand miles (kilometers) in diameter, within which the eclipse is seen as partial. These facts mean that if you are situated on the path of totality, you begin by seeing a long, ever-deepening partial eclipse as the zone of partiality envelops you. Then the zone of totality reaches your location and, if conditions are ideal, can take as much as seven or, in principle, nearly eight minutes to traverse it; during this time the bright disk of the sun is completely hidden from view and you see the corona. Suddenly the bright surface of the sun breaks through again; totality is past, and you are again in the zone of partiality. If you had been a few kilometers away, totality would have missed you altogether and you would have seen only a partial eclipse.

All these effects depend, of course, on the moon being able to cover the whole sun as seen from earth. The moon is only barely big enough to do so; moreover, its distance from the earth varies, and when it is farther away than average (and therefore looks smaller) it can indeed fail to cover the whole sun. What happens then is that there is no zone of totality; an observer in the exact center of the moon's shadow sees the moon centered on the sun but not quite big enough to cover it, so that a ring-like area of the sun's surface remains visible (Fig. 2.2). This phenomenon is called an *annular eclipse* (from Latin *annulus*, "ring"). (See Plate 12.) An annular eclipse will be visible from the United States on May 10, 1994.

Figure 2.3 shows the approximate paths of all total and annular solar eclipses through 1999.

Eye protection

When you view any partial or annular eclipse, or the partial phases of a total eclipse, you are looking at the bright surface of the sun (the *photosphere*) – which is so bright that looking at it through a camera or telescope without proper eye protection can permanently damage your eyes.

> **Never look at the sun through *any* optical instrument unless you are sure it is equipped with filters that make solar viewing safe.**

Obeying this warning is a matter of know-how, not just common sense, for two reasons. First, *this type of eye injury is usually painless;* you may not know at the time that you have suffered any permanent damage. What may happen is that the injured person experiences a sensation of "dazzle" exactly like the normal (and harmless) afterimage from a photographer's

Fig. 2.1. Solar eclipse configurations in space.

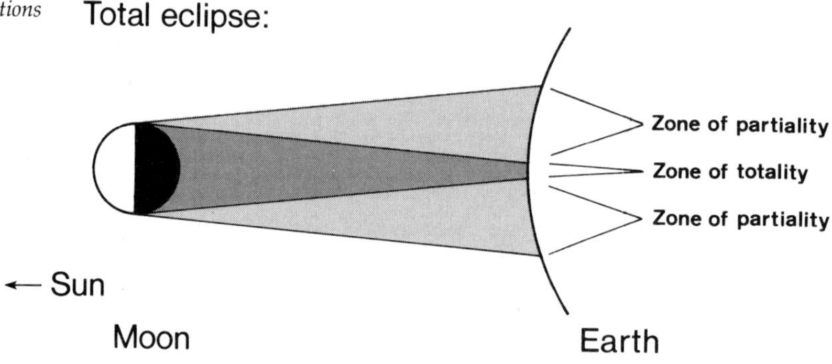

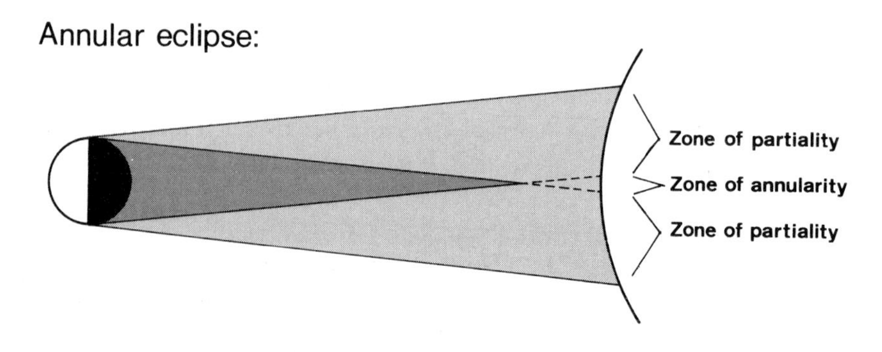

Fig. 2.2. Solar eclipses as seen from earth. Solar filters must be used for eye protection except for the single, central image of totality in the top row.

Total eclipse:

Annular eclipse:

Any eclipse viewed from outside the path of totality or annularity:

Solar eclipses 1994-1999

Total
Annular

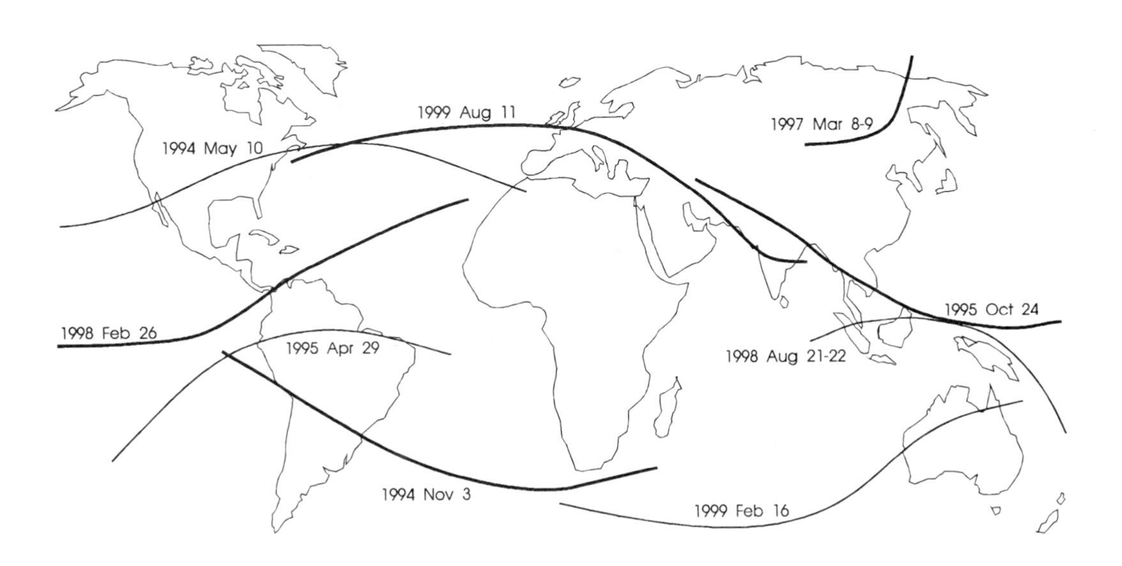

Fig. 2.3. Solar eclipses, 1994–1999.

flash, except that, instead of disappearing after a few minutes, it gets worse and within a few days develops into a permanent blind spot.

Second, and perhaps more important, the fact that the sun looks comfortably dim through a particular filter does not prove that the invisible wavelengths are being reduced to a safe level. The reason is that the eye cannot see all the wavelengths present in sunlight (light at wavelengths too short to see is called *ultraviolet*, and light at wavelengths too long to see, *infrared*). Photographic neutral-density filters are notorious for transmitting dangerous amounts of infrared; more about this later. Smoked glass may not be uniform. Filters are safe only if instrumental tests have shown that they cut infrared and ultraviolet radiation by at least the same amount as visible light and provide uniform attenuation.

Around the time of a major solar eclipse, the news media often quote supposedly authoritative sources as saying that no type of filter is safe for direct viewing of the sun. Taken literally, this is not quite true; several types of filters (to be discussed below) have been shown to be safe when used properly, and astronomers have been using them for years without injury. What the authorities mean is that no type of filter is foolproof enough to be used by complete novices without expert supervision. Safe solar viewing requires more than the average amount of willingness to follow instructions to the letter: the rules cannot be stretched even slightly.

Remember: even through a filter, **never stare at the sun**. Look for only a second or two, even through a filter, to monitor the progress of partial phases.

When organizing viewing sessions for large groups, remember that people with no understanding of optics may sometimes misuse filters in ways that an experienced photographer would never dream of. At the May 1984 annular eclipse in Georgia, at least one schoolboy poked a hole in a filter in order to view the sun "through" it, and several people thought aluminized Mylar filters were made of aluminum foil. Others used insufficiently dense film filters without realizing that there was anything wrong with them, since they had never seen a filter of the correct density. Further, an annular or deep partial eclipse tempts people to look at the sun directly, without a filter. Even though most of the photosphere is covered, and the glare of the sun is much reduced, the uncovered regions are as hazardous as ever.

Still, the often-heard statement that "the only safe way to watch an eclipse is on television" is clearly an exaggeration: it is like saying that the only safe way to see the Grand Canyon is on a picture postcard (because you might fall in if you visit it in person). Eclipse watching can be done either dangerously or safely; it would be unfortunate to miss a spectacle of nature because of unrealistic fears.

Fig. 2.4. Shadow of Michael Covington holding a piece of paper with a ¼-inch hole in it during a partial solar eclipse. The image of the hole takes on the shape of the partially eclipsed sun. (Melody Covington)

Viewing projected images

In any event, you don't have to look directly at the sun at all when it is partially eclipsed or during the annular phase; you can project the sun's image onto a screen and take pictures of the screen. The possibility of eye injury is then eliminated. The simplest instrument for projecting an image of the sun is a pinhole – either a hole punched in a card perhaps 1/16-inch (a millimeter) across and used to cast an image on another card about 8 inches (20 cm) away, or a larger hole, casting a larger image on a larger screen. The hole-to-screen distance should be about 100 to 500 times the diameter of the hole in either case.

The principle involved is that, under such conditions, the spot of light that comes through the hole takes on the shape of the light source, not the hole. (You can demonstrate this fact by using square or triangular holes with the round, uneclipsed sun; if the screen is far enough from the hole, the image will be round.) If the hole is too big or too close to the screen, its shape influences the shape of the image, while if it is too small or too far away, the image is too dim and may be blurred by diffraction. A ¼-inch (7-mm) hole at 6 ft (2 meters), or a 1-inch (25-mm) hole at 24 ft (7 meters), is about right (Fig. 2.4). You can even observe a partial eclipse without any equipment, by making a tiny aperture with your thumb and the base of your index finger, and looking at its shape in the shadow on the ground.

The image that you get with pinhole projection is neither very sharp nor very bright, and you can get much better results by doing sun projection with a telescope (Fig. 2.5). Aim any telescope – or a pair of binoculars with one of the front lenscaps left on – at the sun, taking care not to look through it; hold a white card 4 or 6 inches (10 or 15 centimeters) behind the eyepiece, and focus until you get a sharp image of the sun (the edge should look crisp, and it should normally be easy to see one or two sunspots). This technique will generally give a good solar image, of adequate brightness, at least twice as large as the telescope objective; for instance, a 6-inch (15-cm) telescope will give a bright image a foot (30 cm) across. The screen on which the image is projected will of course have to be shaded from direct sunlight; a convenient way to do this is to use a telescope that projects its image at a right angle, such as a Newtonian, or a refractor with a star diagonal, and put the screen at right angles to the sun. By doing this it is easy to display an 8-inch (20-cm) image on a slide projector screen, where several people can watch it. If you can project into a darkened room or tent, you can work with a dim but gigantic image a meter or so across.

Three precautions are in order. First, and most obviously, *make sure that no one tries to look into the eyepiece*. A good way to convince onlookers, especially young ones, that doing so would be unwise is to hold a piece of paper in the concentrated beam of sunlight emerging from the eyepiece; it will catch fire immediately. I have heard of a British astronomer who lights his pipe this way.

Second, note that the concentrated heat can damage eyepieces. Use an eyepiece that contains no cemented elements (such as a Ramsden or Huygenian), or an old or cheap eyepiece that you consider expendable. Allow it to cool down periodically by covering the front of the telescope or aiming it well away from the sun. (Do not aim it only *slightly* away from the sun; the concentrated light and heat would then fall on the inside of the tube, possibly setting fire to it.) It is usually a good idea to mask the telescope's aperture to 3 inches (7 cm) or so; a square mask may be easier to make than a round one and is just as good.

Third, cover the finderscope so that you don't have any unpleasant encounters with the small solar image that it projects.

Once you have the sun's image projected onto a screen, of whatever size, it is a straightforward matter to take a picture of the screen with the image on it. Use an exposure meter as usual, whatever kind of film is handy, and remember that spots on the screen will not be distinguishable, in the picture, from spots on the sun.

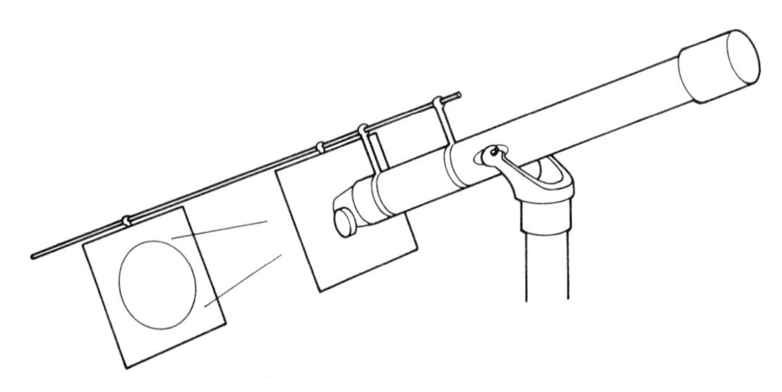

Fig. 2.5. *Projecting an image of the sun with a telescope. The first card, around the tube, casts a shadow so that the image on the second card is easy to see.*

Table 2.1. *Three ways of measuring the light transmission of filters*

Per cent transmission	Filter factor	Logarithmic density (*D*)
50	×2	0.3
25	×4	0.6
10	×10	1.0
1	×100	2.0
0.1	×1,000	3.0
0.01	×10,000	4.0
0.001	×100,000	5.0
0.0001	×1,000,000	6.0

The density value is usually given as ND (neutral density). "Neutral" means that it cuts all colors equally.

Sometimes natural sky filtering will reduce the sun's intensity drastically (Plate 10). The 1984 eclipse was annular but, unusually, so close to total that some unfiltered photos could be taken (Fig. 2.6 and Plate 11).

Photography through filters

Although projection setups are ideal for group viewing, they leave much to be desired as far as photography is concerned. The alternative, then, is to use a telephoto lens or telescope in combination with a protective filter. One of the most important characteristics of a sun filter is, of course, the extent to which it reduces the intensity of the light. As Table 2.1 shows, there are three ways of measuring this: as a percentage transmission, as a filter factor, and as a logarithmic density. These are related by the formulae:

filter factor = 100/percentage transmission
logarithmic density = $\log_{10}$ (filter factor)

Note that when two filters are combined, the filter factors multiply, but the logarithmic densities add; this is the main convenience of using logarithmic densities. For example, a ×2 neutral density filter and a ×10 filter used together have a combined filter factor of 20 (that is, 2 × 10); expressing the same thing logarithmically, filters with densities of 1.0 and 0.3, used together, have a combined density of 1.3. Be sure not to confuse filter factors with logarithmic densities; a ×4 neutral density filter is a medium gray, while one with a density of 4.0 is so dark that nothing except the sun or a very bright light can be seen through it.

The surface of the sun is about 300,000 times as bright as an ordinary sunlit scene on earth; this means that in order to reduce its brightness to a comfortable level, you need a filter whose logarithmic density is about 5.5

Table 2.2. *Safe and unsafe sun filters*

Safe	Metallic glass or film filters designed for solar viewing and used as directed (*best*)
	Two or three layers of fully exposed and developed conventional black-and-white film
	No. 14 welder's glass
Unsafe	Photographic neutral density filters of any density
	All other combinations of photographic filters, including crossed polarizers
	Filters made of color film
	Filters made of chromogenic ('silverless') black-and-white film (such as Ilford XP2)
	Smoked glass
	Any filter through which you can see things other than the sun and very bright electric lights
	Any filter located near the telescope eyepiece, unless used with an unsilvered mirror or Herschel wedge
	Any filter not *known* to be safe

(= $\log_{10} 300{,}000$). In practice, densities of about 4.0 to 6.0 are used (the lower densities on high-magnification telescopes and slow lenses that give an intrinsically dimmer image). The important thing is not so much the visual density as the density in the infrared and ultraviolet, which you can't see. A thorough set of safety tests for sun filters has been conducted by Dr. B. Ralph Chou of the Optometry School at the University of Waterloo in Ontario. ('Safe Solar Filters,' *Sky and Telescope*, August 1981, pp. 119–21; 'Protective Filters for Solar Observation,' *Journal of the Royal Astronomical Society of Canada*, vol. 75, pp. 36–45, 1981.) Table 2.2 summarizes his results.

In particular, note that photographic neutral density filters (Wratten #96 or the like) are *not* safe to look through. This is a pity, since they are made in a wide range of accurately regulated densities. They are perfectly satisfactory for solar photography, of course, provided you can find some way of aiming and focusing your camera without looking at the sun through it.

The cheapest way to get a safe sun filter is to take a roll of conventional black-and-white film, unroll it in daylight or full room light so that it is fully exposed, then develop and fix it in the normal manner and use two or three layers of the resulting black film as a sun filter. (If three layers are too opaque to allow you to see the sun, then try two layers. For very dense film such as x-ray film, even one layer may do.) A 120-size roll of Verichrome Pan, which is very inexpensive, makes a piece of black film about 2¼ × 30 inches (55 × 750 mm); Kodak T-Max 100 and Technical Pan and Ilford Pan F and FP4 are also suitable. (You can develop the film in full room light, since it's fully exposed; or you can have it developed commercially.) Do not use color film or chromogenic ("silverless") black-and-white film.

The trouble with filters made of exposed film is that they are not very good optically; they don't give sharp images with lenses longer than about 200 mm in focal length. The same is true of welder's glass. But excellent optical quality combined with first-rate eye protection can be obtained by using a filter that consists of thin coatings of metal on glass or plastic. Of these, one of the best is the "Solar Skreen®" marketed by Roger W. Tuthill (Box 1086, Mountainside, NJ 07092, Tel. 800 223-1063 or 908 232-1786); it consists of two layers of Mylar plastic each of which has an aluminum coating on both sides, and it mounts in front of the telescope. Its only disadvantage is that it makes the sun look a peculiar electric blue color, but this is no real loss since the sun is a rather colorless object; for more realistic-looking color photographs, a #12 or #15 yellow filter can be added.

Somewhat more expensive are glass filters with a reflective metal coating. They give a pleasing orange tinge to the sun. Such filters can be ordered cut to the size of your lens, and with a mounting ring, from Thousand Oaks Optical (Box 248098, Farmington, MI 48332-8098, Tel. 313 353-6825). Their filters are made to transmit 0.001%, that is, 1/100,000 of the incident light, which has 5 zeroes and is therefore ND5. Actual samples may be ND4.7 to 5.2, which is satisfactory. For photographers wanting shorter exposures, they also have a line of ND4 filters.

The importance of placing the sun filter in front of the telescope cannot be overstated. *A sun filter placed at or near the eyepiece is not safe*; the concentrated heat of the sun can crack and melt it, with disastrous results. This is true even if the telescope is quite small, since virtually all of the light that does not go into your eye has to be converted to heat within the filter. (Even giving the filter a shiny, reflective surface – as was commonly done a century ago – does not reduce heat absorption to a safe level.) The only safe way to use an eyepiece-mounted sun filter is to reflect the light off an unsilvered glass surface (a Herschel wedge) first – and, even then, you have to deal with image degradation resulting from heat waves in the air inside the telescope. It's better to keep full sunlight out of the telescope altogether by using a front-mounted filter.

Apart from filtration, the optical requirements for photographing the sun are the same as for photographing the moon – in fact, your eclipse-photography setup, minus the filter, can be tested on the full moon two weeks before the eclipse. The image sizes for the sun and the moon are exactly the same, and the time limits for exposures in Table 3.3 (Chapter 3) are equally applicable – though with the sun it is easy to keep exposures short by choosing filters of appropriate density.

Moreover, partial solar eclipses can be photographed on practically any kind of film. The exposure needed depends on the filtration used and can be determined in advance of the eclipse by practicing on the uneclipsed sun. In past years it was customary to take advantage of the sun's brightness and use extremely slow film for solar photography, but this strikes me as a bad idea – if your filtration is so light that you can photograph the sun on films significantly slower than you would use without a filter for photographing the moon, then it probably isn't safe to look into the eyepiece.

There is one additional precaution that may not occur to you in advance, since the need for it arises only in the daytime: if you're using an afocal

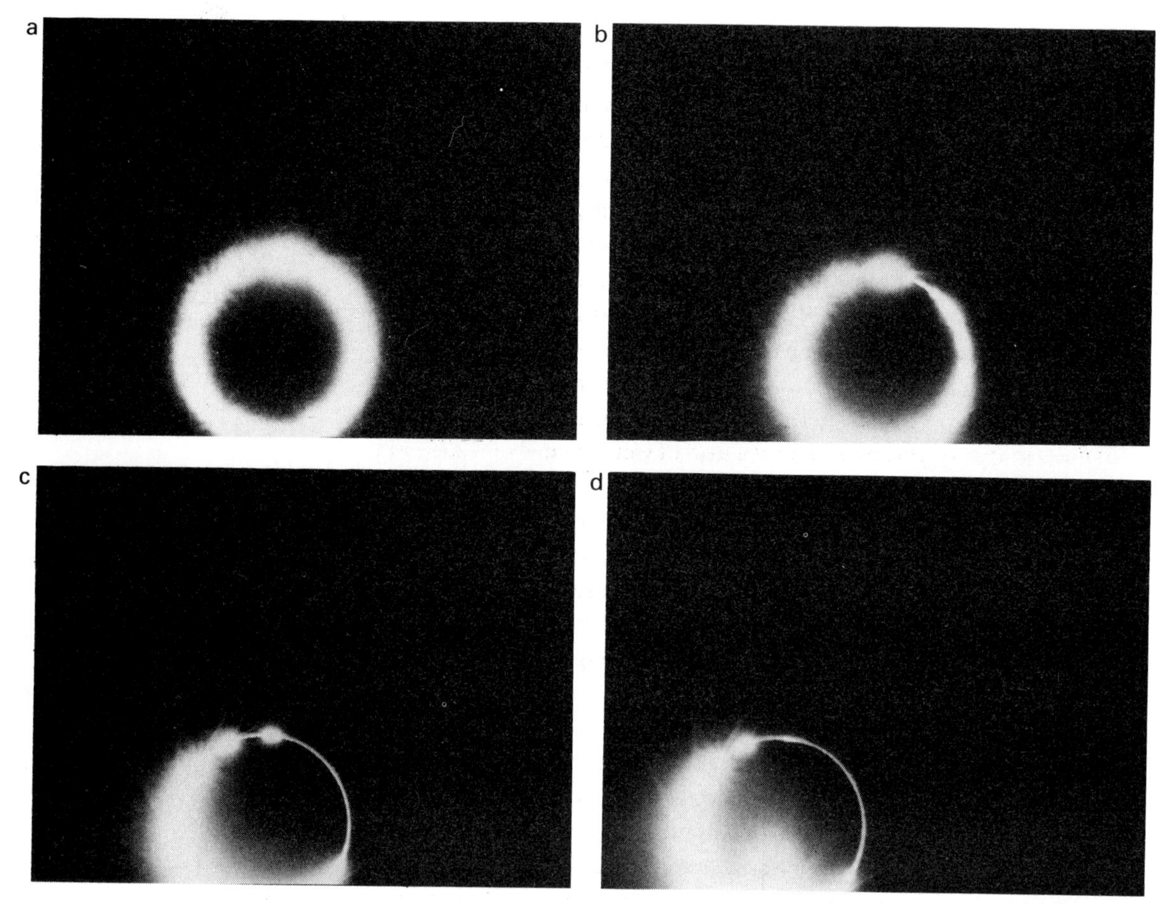

***Fig. 2.6.** The annular solar eclipse of May 30, 1984, photographed from Pendergrass, Georgia, by Melody Covington. The moon covered 99.9% of the sun, much more than most annular eclipses. The 1994 eclipse will have far more of the sun remaining in the annulus (6% instead of 0.1%) so solar filters will have to be used throughout the eclipse, unlike the case here. (a) Mid-eclipse: the moon (black) is encircled by the greatly overexposed solar photosphere. (b) The moon has moved aside slightly, creating Baily's beads at the top. (c) and (d) Toward the upper right, the photosphere is completely covered and the chromosphere is visible. The thick spot in the middle of it is a prominence. Exposures about 1 second apart with a 400-mm lens at f/32, 1/1000 second, on Ektachrome 200, subsequently enlarged in slide duplicator. Because no filter was used, it was not possible to look through the viewfinder, and an attempt to aim the camera by its shadow was not wholly successful – the sun drifted out of the field.*

setup (p. 54) to photograph the sun, stray light has to be kept out of the space between the eyepiece and the camera lens. A convenient way to do this is to drape a piece of black cloth loosely around the camera and telescope after positioning them.

Totality – the corona

The totally eclipsed sun is at once an easy and a difficult photographic target – easy, in that many different equipment configurations, films, and exposures can give pleasing results, but difficult, in that no photograph can record all the coronal structure and color visible to the eye. The reason for this is that the corona is much brighter nearer the center than at the extremities. Your eye adjusts to this brightness variation automatically, but photographic film does not; if you expose for the outer corona, you overexpose the inner regions, whereas if you expose for the inner corona, you lose the outer parts completely (Fig. 2.10). This makes it difficult to photograph more than a small amount of the streamer-like structure that is so striking visually; but at the same time it ensures that any exposure within quite a wide range will be right for *some* part of the corona.

To photograph the corona you need a field of view somewhat wider, and hence a focal length somewhat shorter, than for the partial phases of the eclipse. The useful effective focal lengths with 35-mm film range from about 1500 mm, which covers about twice the diameter of the solar disk, down to 200 mm or so; even a 50-mm "normal" lens can be useful in recording the outermost parts of the corona. Suggested exposures are given in Appendix B. It goes without saying that, as no light from the bright photosphere can reach you, no filters are needed for viewing or photographing the sun during totality.

The best way to tackle the problem of the corona's brightness range is to use a color negative film and make, or get someone to make, a dodged print in which the relative density of the inner and outer parts of the picture is adjusted by hand. It is possible to construct a special dodging mask to get the right amount of correction more or less automatically; for two ways of doing so, see "Some hints for photographers of total solar eclipses," *Sky and Telescope*, May 1973, pp. 322–6. The same article describes a way to use an occulting disk inside the camera to establish a density gradient with much the same effect.

Among color print films, the Kodacolor Gold series is noted for its exceptional exposure latitude. Color slide films can also give good results; choose a relatively fast film (like Kodachrome 200 or Ektachrome 200) with good exposure latitude, and stay away from high-contrast materials. Nowadays, few people photograph the corona in black and white, but if you

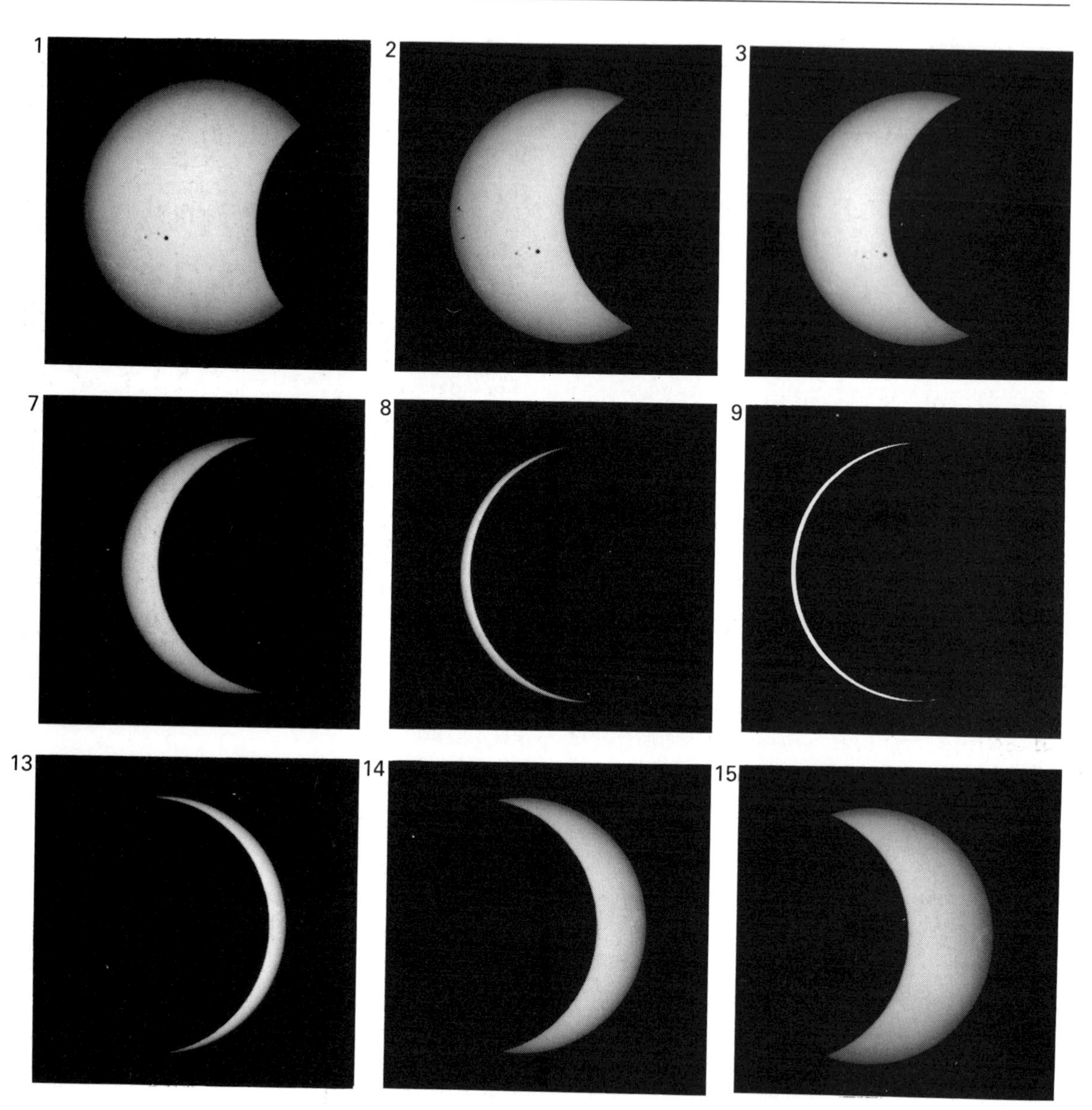

Fig. 2.7. *The annular eclipse of May 30, 1984, photographed from Pendergrass, Georgia, with a 12.5-cm (5-inch) f/10 Schmidt–Cassegrain telescope and full-aperture Solar Skreen filter. Each exposure is 1/30 second on Kodak Technical Pan Film 2415 developed 6 minutes in HC-110(D) at 20°C (68°F). At mid-eclipse, almost all the photosphere (99.9%) was covered. (By Michael Covington)*

(Fig. 2.7. cont.)

choose to do so, develop the film to slightly lower than normal contrast in a developer such as Microdol-X.

The light of the corona is partly polarized, and it is interesting to take several pictures through a polarizing filter, changing the orientation of the polarizer for each exposure.

See Figures 2.6 and 2.9 and Plate 9.

Beads and prominences

But the corona is not the only thing to take pictures of. At the beginning of totality, the thin silver photosphere that has been visible suddenly breaks up into a number of disconnected spots, called *Baily's beads,* which consist of light coming through the spaces between lunar mountains. Within a few seconds, all of the spots disappear except one. What remains is called the *diamond-ring effect* – the single gleaming bright spot together with the ringlike appearance of the inner corona look rather like a diamond ring in a jeweler's advertisement. (See Plate 8.)

A moment later, the photosphere is hidden completely and the corona has come fully into view. But where there was a thin sliver of photosphere half a minute ago, there is now a thin, glowing, reddish sliver of something else – the *chromosphere* a layer of ionized gas that lies between the photo-

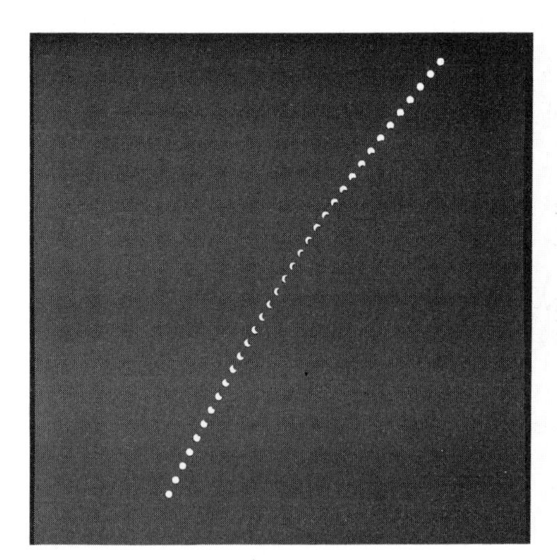

Fig. 2.8. *Multiple exposures at 5-minute intervals of the solar eclipse of July 31, 1981, taken with a Mamiya C330 camera on a fixed tripod with a 100-mm lens at f/16 and a filter of logarithmic density 4.0. Each exposure was 1/60 second on Ektachrome 64. (Akira Fujii)*

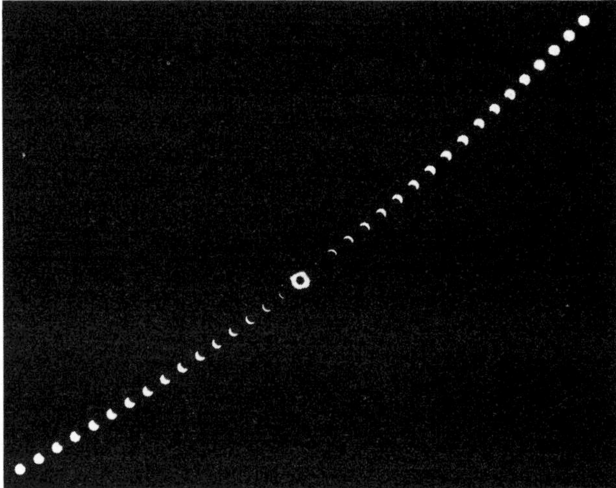

Fig. 2.9. *The July 11, 1991, total solar eclipse, photographed in time lapse from Baja California, also by the distinguished Japanese amateur astronomer Akira Fujii. He used a Mamiya Press camera with a 100-mm lens; the interval between exposures was 5 minutes see also Plate 9.*

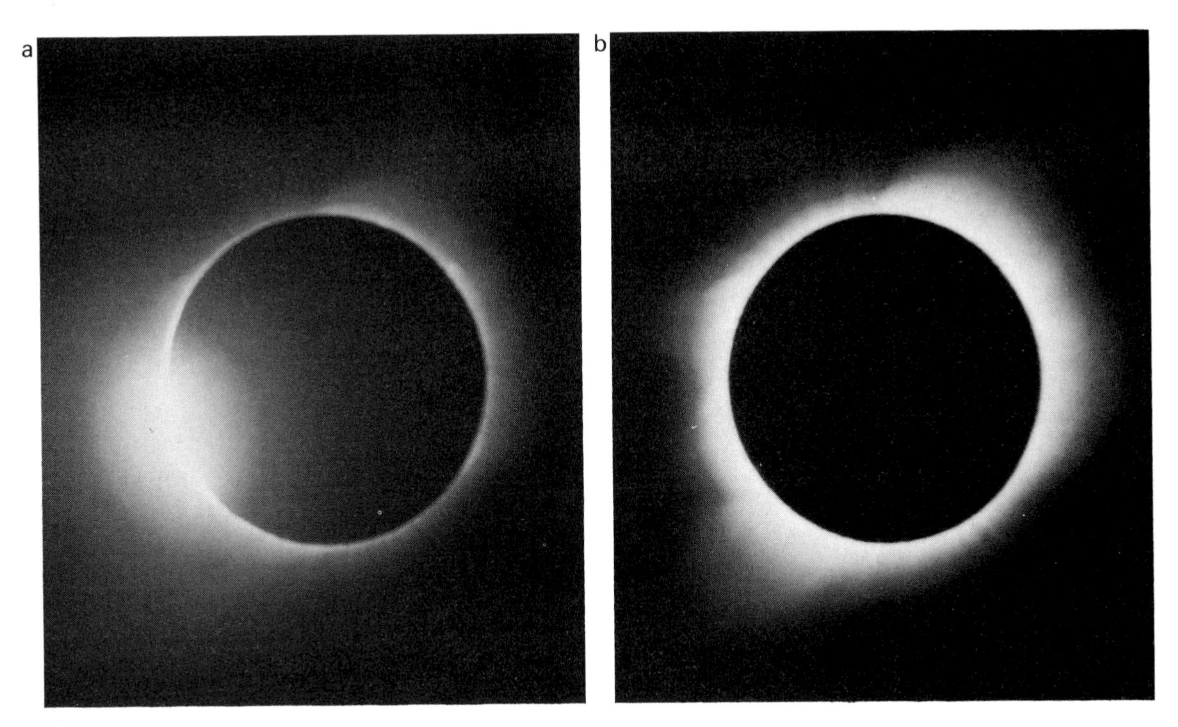

Fig. 2.10. Different exposures show different amounts of corona. (a) ¼-second on ISO-100 Kodak Plus-X Film with a 500-mm f/8 Nikkor lens on a Nikon F2as. The diamond ring also shows. (b) 2 seconds, same setup. Both were taken at the 1984 eclipse in Papua New Guinea. (Jay M. Pasachoff)

sphere and the corona. ("Chromosphere" simply means "color sphere.") The chromosphere will likewise disappear from view within a few seconds, so the opportunity to capture it on film must be seized quickly. Finally, there are the *prominences*, great fountains of gas that glow red like the chromosphere but extend upward into the inner corona, usually high enough to remain visible throughout totality. (They appear reddish because they emit strongly in the red color typical of hydrogen gas.) At the end of totality the chromosphere, the diamond-ring effect, and Baily's beads reappear in succession on the opposite edge of the moon.

The tables in Appendix B give suggested exposures for prominences; these exposures should do equally well for the chromosphere. Exposures for Baily's beads and the diamond-ring effect are hard to estimate in advance; use a slightly shorter exposure than for prominences, and hope for the best. Such photographs are usually only good when taken with a telephoto lens of at least 300 and preferably 500 mm. The camera must be mounted on a tripod if possible. Remember that the best time to photograph Baily's beads and the diamond-ring effect is usually at the *beginning* of totality, while you can safely start looking into the eyepiece and then continue doing so. When you see the first spot of photosphere at the end, take your eye away from the viewfinder *immediately*, but you can continue taking photographs even though you are no longer looking through the eyepiece. Within a second or two, put the sun filters back on, or simply point the camera away from the sun.

Don't confuse *prominences* with *solar flares*. The latter are much hotter, more explosive, and shorter lived. We don't usually see them during eclipses.

Shadow bands

There are also interesting things to see on the ground. Beginning about a minute before totality, and continuing until totality actually begins, the ground is covered with fast-moving parallel *shadow bands* a few centimeters wide, which reappear for an equal length of time at the end of totality. Shadow bands probably result from irregular refraction of the crescent sun's image by the earth's atmosphere. Since they are terrestrial rather than solar phenomena, astronomers are not usually interested in them.

Shadow bands are exceptionally hard to photograph because of their rapid motion and low contrast (there may be only a 1% difference in brightness between the bright and dark bands) and because the overall light level is constantly changing. Shadow bands are most easily seen and photographed on a bright white surface – ideally, a slide projector screen or a white sheet or artist's board – which the sun's rays are striking as directly as possible. Use a fast film with good exposure latitude, such as Kodak Tri-X Pan or Ilford XP2 (both ISO 400 black-and-white films), and make the prints on high-contrast paper or increase the contrast by repeated copying. (Do not use a high-contrast film for the original exposure; it would not give enough exposure latitude.) You could also try a fast color film like Kodak's Kodacolor Gold 400 or a video camera.

The correct exposure for shadow bands is hard to predict; a rough guess is 1/250 second at *f*/4 (or 1/500 at *f*/2) on ISO (ASA) 400 film. An automatic-exposure camera set for aperture-priority, especially one that measures the light level throughout the exposure, can prove very helpful, since it can follow the changing light levels automatically; set the lens at its widest opening and hope for the best. The scientific value of pictures taken with an automatic-exposure camera is increased if there is some way of determining what shutter speed the automatic mechanism has given you; one way of doing this is to include in the picture an object moving at a known speed, such as a rotating disk, and measure how much of a blur it leaves on the film. Some cameras are available that imprint such information on the film. The shadow bands themselves move too fast to be visible in exposures of more than 1/125 second, and much shorter exposures, of the order of 1/1,000 second, are preferable.

Horizon views

Interesting effects are visible in the sky and on the horizon during totality. A 10-second fixed-tripod exposure of a large area of the sky will record the outer corona, a few bright stars, and any planets or comets that may be visible. (It is quite possible for a bright comet to be discovered during an eclipse, having been too close to the sun to be visible under ordinary conditions.) The horizon may appear orange or maroon in color, a 360° effect much like a sunset in all directions. Before and after totality, the rapid motion of the zone of totality – which you can see coming at or going away from you out of the distance as a cone of gathering darkness – is awe-inspiring.

Session planning

Totality is, of course, short; you are fortunate if you have three or four minutes in which to take all your photographs. This means that things have to be planned carefully and practiced in advance; make several "dry runs," without film, of the whole sequence of activities you plan to carry out during totality. It is best to plan on using only about 70% of the time you will have available; things will inevitably go wrong and slow you down.

Wearing sunglasses for the last half hour before totality will help your eyes to dark-adapt. Some people even wear an eyepatch on one eye. A tape recorder, playing a tape you have prepared in advance, can be useful for timing; you can prepare a verbal countdown so that you don't have to look at your watch to find out how many seconds of totality are left. This tape recorder can be supplemented with another tape recorder to record your comments so that no time is lost writing things down. A camera with motorized film advance can save some time in making the exposures. (If you use one – or even if you don't – be sure not to use up all the exposures prematurely.)

But don't get so busy taking pictures that you neglect the visual aspects of the eclipse. The pictures you take will probably look very much like the pictures hundreds of other people are taking at the same time, but neither your pictures nor theirs will capture the visual glory of the delicate coronal streamers or the feel of the sudden onrush of the moon's shadow. You can look at photographs any time; you may see totality "live" for only a few minutes in your whole life.

CHAPTER THREE

CAMERAS, LENSES, FILMS, AND TELESCOPES

Cameras and lenses

The ideal camera for eclipse photography is a 35-mm single-lens reflex (SLR). (We discuss video cameras in Chapter 7.) Auto exposure is a useful feature if it works well at low light levels; autofocus is useless or worse. A mirror lock button, to keep the mirror from introducing vibration, is helpful but not essential. Even if your camera is not an SLR, you can couple it to a telescope using the afocal method (p. 54).

To photograph the sun or moon you need a long telephoto lens or a telescope of some kind; your camera's normal lens by itself won't give a large enough image to show any detail (Fig. 3.1). The size of the image of the sun or moon on the film depends on the focal length of the lens, as expressed by the formula:

image size (on film) = focal length/110

where the focal length is expressed in the same units as the image size (normally millimeters). As an approximation, just remember that the image size is a bit less than 1/100 of the focal length. Importantly, images of the sun and of the moon are about the same size.

Table 3.1 shows the image size you'll get with lenses of various focal lengths, both on the film and on a ×15 enlargement (the highest degree of enlargement that is practical with most films and lenses – about the equivalent of a 16 × 20 inch print from a 35-mm negative, an 8 × 10 from a 110 negative, or a 30 × 30 from a 120 negative, if you were to enlarge the entire picture area). Table 3.2 shows the field of view for various lenses when used with a 35-mm camera.

Let's suppose you're working with a 35-mm camera. If you use a 50-mm lens, the image of the sun (through a filter) or moon on the film will be about half a millimeter in diameter, and even with maximum enlargement, you'll only get an image about 7 mm (about a quarter of an inch) across on the print

Table 3.1. *Size of sun (disk, no corona) or moon image with various focal lengths*

Focal length (mm)	Size of image on film (mm)	Approximate size on ×15 enlargement (mm)	(inches)
28	0.25	3.8	1/8
50	0.45	6.8	1/4
100	0.91	14	1/2
200	1.8	27	1
300	2.7	41	1 5/8
400	3.6	54	2 1/8
500	4.5	68	2 5/8
600	5.4	81	3 1/8
750	6.8	102	4
900	8.2	123	4 3/4
1,000	9.1	136	5 3/8
1,200	10.9	164	6 3/8
1,500	13.6	205	8
2,000	18.2	273	10 3/4

Fig. 3.1. *The moon over Mt. Wilson Observatory as photographed by Jim Baumgardt. Because a 50-mm lens was used, the image of the moon is quite small.*

Table 3.2. *Field of view of different lenses with 35-mm cameras*

Focal length (mm)	Angular diagonal on 35-mm film	Field of view
28	85°	73° × 50°
50	50°	41° × 28°
100	25°	20° × 14°
200	12°	10° × 7°
300	8°	7° × 5°
400	6°	5° × 4°
500	5°	4° × 3°
600	4°	3.4° × 2.3°
750	3.3°	2.8° × 1.9°
900	2.8°	2.4° × 1.6°
1,000	2.5°	2.1° × 1.4°
1,200	2.0°	1.7° × 1.1°
1,500	1.7°	1.4° × 1.0°
2,000	1.2°	1.0° × 0.7°

Note: The angular diameter of the sun and moon is ½°. The sun with 1 solar diameter of corona all around it is 1½° across.

– far too tiny to show any detail. So you switch to your longest telephoto lens, a 400-mm. This gives you a 3.6-millimeter image of the sun or moon that you can enlarge to about 55 mm (just over 2 inches) – big enough to show the lunar maria (''seas'') and the face of the man in the moon, and perhaps some of the larger craters (Figs. 3.2 and 3.3) or sunspots.

If you then add a ×2 teleconverter, the effective focal length will be 800 mm, which is well within the useful range for photographing lunar or solar detail; you can get an enlarged image of about 100 mm (4 inches) diameter, and, if your lenses are sharp enough and your tripod is steady enough, you may begin to see craters and mountains on the moon, and the largest sunspots should be visible on the sun. A lens whose focal length is 600 mm or more by itself, without a converter, would of course be even better. Figure 3.4 compares the effects of focal lengths of 600, 1,500 and 2,500 mm; a focal length of 1,000 mm is quite ample for solar and lunar photography on 35-mm film.

Here are some important points to remember in photographing the sun or moon through a telephoto lens:

1 Always use a tripod. You need as sharp an image as your camera can give, and you won't get maximum sharpness unless your camera is mounted on a *very steady* tripod. Use a cable release or self-timer to prevent the camera vibrating when you click the shutter. Needless to

Fig. 3.2. Through a long telephoto lens, the moon shows considerable surface detail. 400-mm Soligor lens at f/6.3, 1/500 second on Ektachrome 200. (Michael Covington)

Fig. 3.3. The moon as photographed through a 400-mm telephoto lens, 1/500 second at f/6.3 on Kodak Technical Pan Film 2415 developed 8 minutes in HC-110(D) at 20°C (68°F) and enlarged about ×18. The crater Tycho is prominent in the white area at the top. (Melody and Michael Covington)

say, it is almost never possible to take acceptable solar or lunar photographs with a hand-held camera, even when the exposure is quite short.

2 Set the lens at about *f*/5.6 or *f*/8 if possible, particularly if you are using a teleconverter. Most lenses are sharpest in this range. The lowest *f*-stop is usually the least sharp.

3 Check the focus. If you're using a lens longer than 135 mm, or a teleconverter with any lens, you can't just set the focus to infinity and snap away; the lens may not actually be at infinity focus. The teleconverter introduces some errors of its own and magnifies any that are already present. The focus must always be checked at the viewfinder. To do so, rotate the focus cylinder back and forth from one side of the correct focus to the other, to help you zero in accurately. Turn off the autofocus before totality; it sometimes "hunts" and may prevent you from photographing when you want to do so.

4 Remember that teleconverters multiply the *f*-ratio as well as the focal length. For example, a 100-mm lens set at *f*/8 becomes 200 mm at *f*/16 with a ×2 converter, or 300 mm at *f*/24 with a ×3. This has to be taken into account in calculating the exposure.

5 It's very hard to see or photograph craters on the full moon because

the light is striking it so flatly. To capture the roughness of the moon's surface, photograph it when it is lit from the side, particularly at crescent or quarter phases.

Determining exposures

We said earlier that you can sometimes use auto exposure when taking pictures of the sun (through a filter) or of the moon. This is true – but only if your camera has through-the-lens metering, and only if you are dealing with an effective focal length of 2,000 mm or more, so that the sun or moon more or less fills the whole picture. Otherwise, the meter will average the bright sun or moon with the pitch-black background and the sun or moon will be overexposed in the picture. Some cameras give a choice of which metering method to use – matrix, average, or spot. It is best to use the spot choice for the uneclipsed or partially eclipsed sun (through the filter) or for the moon. During totality, it is best to bracket on closeups, though a matrix or averaging light meter would be fine for wide-angle views.

"Bracketing" means to take not only the exposure you measure or calculate but also exposures at adjacent values of exposure time or *f*-ratio. The motto for solar eclipse photography is **bracket widely**. You usually do so by clicking the lens *f*-ratio or the shutter speed dial through adjacent click stops to the normally "correct" exposure.

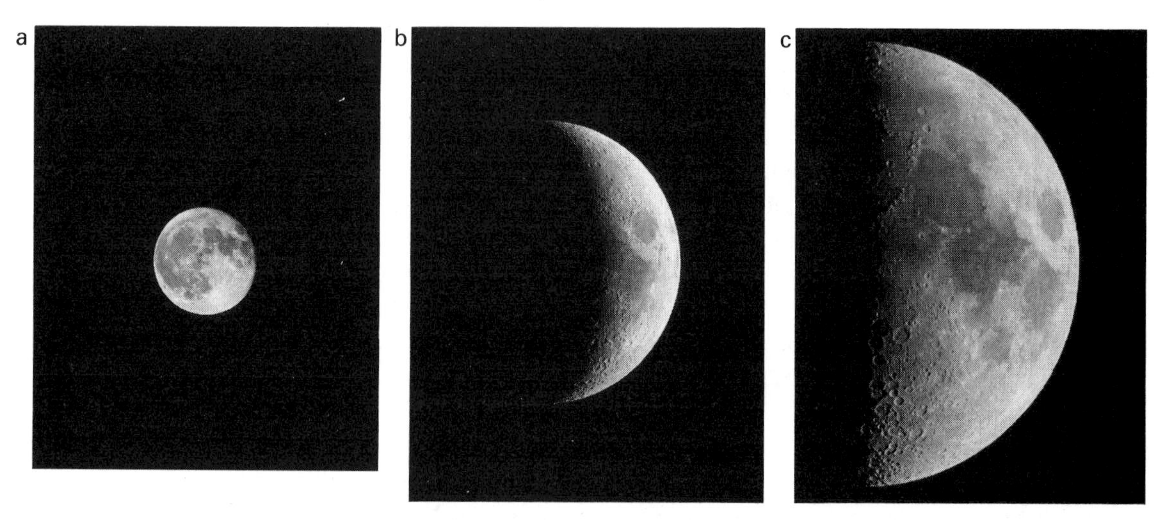

Fig. 3.4. *The moon as photographed with (a) a 600-mm telephoto lens; (b) a 32-cm (12.5-inch) telescope of 1500-mm focal length; (c) a 32-cm telescope of 2500-mm focal length. (Jim Baumgardt)*

Ordinarily, then, you have to determine exposures from tables or by calculation. The basic formula for calculating any exposure is:

$$\text{time (in seconds)} = f^2/(A \times B)$$

where f is the f-stop to which the lens is set, A is the film speed expressed as an ISO (ASA) number, and B is a constant that indicates the brightness of the object being photographed. The uneclipsed full moon has a B value of 200; B values for other situations are given in Appendix B. The B value for the sun through a filter should be determined by experiment.

Here's an example. If you are taking a picture of the full moon at $f/16$ on ISO (ASA) 400 film, plug the values into the formula as follows:

$$(16^2)/(400 \times 200) = 256/80{,}000 = 0.0032 = 1/312 \text{ second}$$

Naturally, your camera won't have a 1/312-second setting; use the nearest that it does have (probably 1/250). The calculated exposures are only approximations; variations in the transparency of the air and other factors can throw them off, so always bracket your exposures.

The same exposure formula as above, with appropriate values of B, applies to almost any picture-taking situation in the sky or on earth. It explains, among other things, why the standard f-numbers marked on most lenses are what they are. According to the formula, the exposure time is proportional, not to the f-ratio itself, but to the square of the f-ratio (f^2). The standard f-numbers and their squares are as follows:

f	1.4	2	2.8	4	5.6	8	11.3	16
f^2	2	4	8	16	32	64	128	256

You can easily see what is happening – whenever you go from one f-stop to the next higher one, you double f^2, and hence you double the required exposure time. The standard shutter speeds (1/1,000, 1/500, 1/250, 1/125, etc.) are likewise arranged so that each one gives twice as much exposure as the one below. Whenever you change the lens opening by a particular number of stops, you can change the shutter speed by the same number of steps in the opposite direction to get an equivalent exposure. A "one-stop" exposure change means that the amount of light or the exposure duration is cut in half or doubled.

There is one other thing to take into account: the earth is rotating, and if your camera or telescope doesn't have a clock drive to counteract its motion, there is a limit to how long an exposure you can make. The situation is the same as with fixed-camera star photography. The formula to use is:

$$\text{longest practical exposure (in seconds)} = 250/F$$

where F is the focal length in millimeters. The moon's declination doesn't have to be taken into account because it doesn't vary enough to have a significant effect. Table 3.3 summarizes the results.

Table 3.3. *Longest exposure that will give sharp images without a clock drive*
These times apply to any celestial object.

Effective focal length range (mm)	For critical work (in seconds)	Where some blur is tolerable (in seconds)
90–180	2	8
180–350	1	4
350–700	1/2	2
700–1,500	1/4	1
1,500–3,000	1/8	1/2
3,000–6,000	1/15	1/4
6,000 and up	1/30	1/8

Choosing a film

As always in photography, your choice of film will be a tradeoff between fine grain and sensitivity. The "faster" the film – the higher the ISO (ASA) number – the shorter the exposure time you can use and so the less image motion from the sun's motion or from vibration. But films of ISO about 100 have finer grain and these give sharper images than films of higher ISO. We have found film of ISO 200 (Kodachrome 200 or Ektachrome 200, both slide films) or even 400 (Kodacolor Gold 400, a print film) to be satisfactory. But films of ISO 800, 1,000, or more, seem to us to give images that are too grainy to be pleasing.

Coupling camera to telescope

Figure 3.5 shows the main types of astronomical telescopes, and Fig. 3.6 shows ways of coupling a camera to a telescope. The most basic one of these is *prime-focus* coupling, in which the telescope itself, without an eyepiece, takes the place of the camera lens. The focal length and *f*-ratio are then simply those of the telescope objective.

Prime-focus photography is feasible with any telescope that has enough *back focus*, i.e., distance between the end of the tube and the position of the image. Figure 3.7 shows that the image needs to be about 2 inches (5 cm) farther out when a camera body is put in place of the eyepiece. Most refractors, Cassegrains, Maksutov–Cassegrains (Questars, for example),

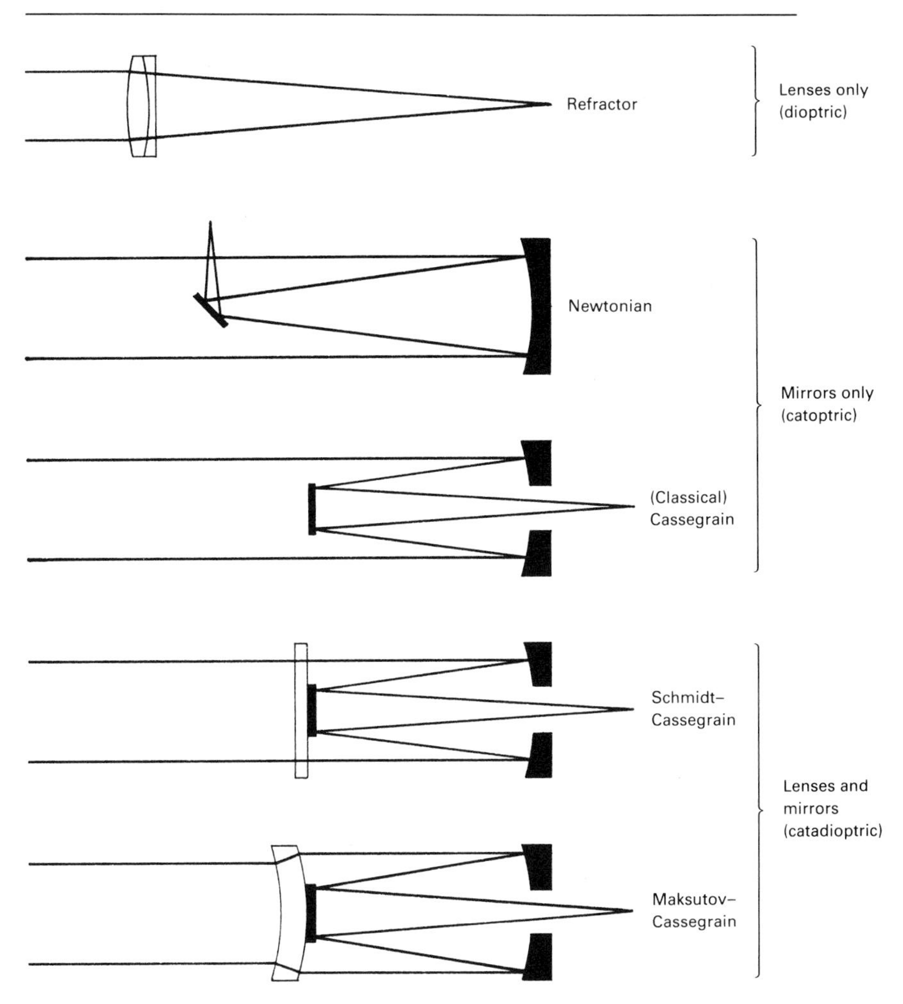

Fig. 3.5. *Five types of telescope. For convenience, the other diagrams in this chapter show the telescope as a refractor, but the principles illustrated are equally applicable to all five types.*

and Schmidt–Cassegrains (many Celestrons and Meades, for example) can manage this; Newtonians cannot. Figure 3.8 shows the exquisite detail that prime-focus photography can produce.

The other types of coupling are covered in detail in *Astrophotography for the Amateur* by Michael Covington (Cambridge University Press, 1991). Positive and negative projection produce an enlarged image; this enlargement is not normally needed when photographing an eclipse, since the image is big enough already. Compression produces a small, bright image at the expense of sharpness; it is used in deep-sky photography and is not suitable for eclipse work.

Afocal coupling

In the afocal method, the camera takes the place of your eye at the eyepiece of the telescope; that is, you simply aim the camera (with its lens in place) into the eyepiece. No special adapter is needed; in fact, the best way to arrange the equipment is to put the camera and the telescope on separate tripods so that neither can transmit vibration to the other (Fig. 3.9). The camera should be equipped with a normal or medium telephoto lens, set to its widest opening. Take care to get the camera lens as close to the eyepiece and as well centered as possible. Use a loose black cloth to block out daylight when photographing the sun through a filter. (Make certain though, that the cloth does not get in the light path.)

If your camera is a single-lens reflex, you'll have no trouble focusing. Start

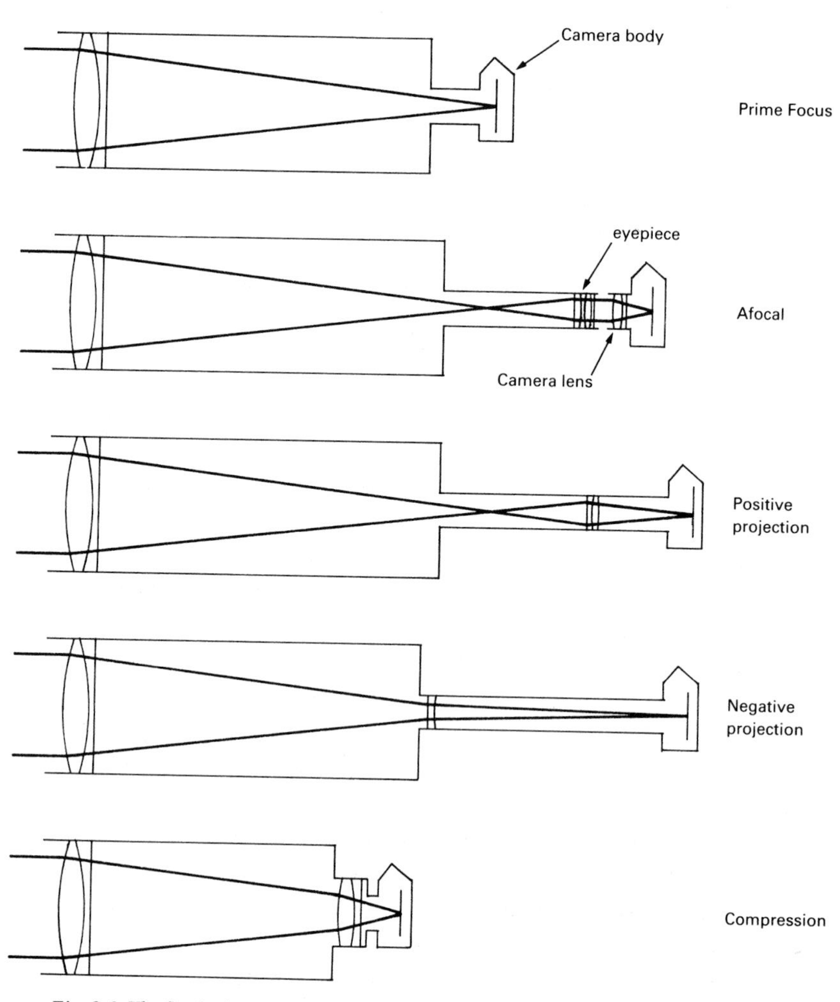

Fig. 3.6. The five basic optical configurations for astrophotography.

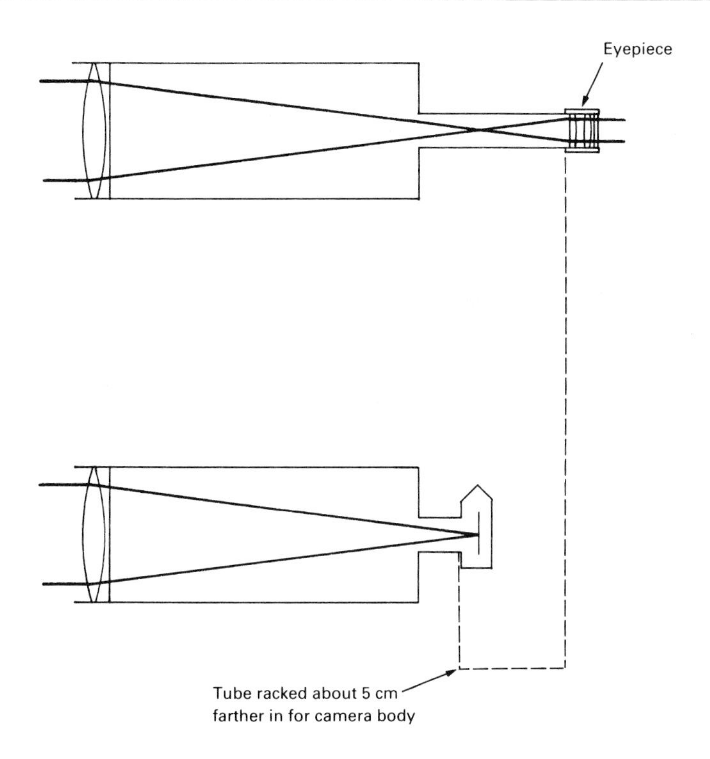

Fig. 3.7. Illustrating the need for back focus.

with the telescope focused for normal viewing and the camera lens set in the middle of its focusing range; look in the viewfinder and focus, using the telescope's focusing knob first and then making fine adjustments by focusing the camera lens. If you happen to have a twin-lens reflex, focus with the viewing lens in front of the eyepiece, then move the camera so as to take the picture through the taking lens.

If your camera doesn't have through-the-lens viewing, you can still focus accurately – but it's a bit more complicated. In addition to the camera and main telescope, you'll need a small hand-held telescope of about 5 to 10 power; your main telescope's finder, or one side of a pair of binoculars, will do fine. First aim the hand-held telescope at the moon, or any other object more than about 500 ft (150 meters) distant, and focus it so that what you see is sharp. Next, aim the small telescope into the eyepiece of the main telescope and focus the main telescope so that you again see a sharp image. Then set your camera to infinity focus (the long-distance end of the focusing scale, usually marked ∞), put the camera in place, and make the exposure.

The hand-held telescope is necessary to reduce the amount of variation that the focusing muscle of your eye can introduce. When you focus your telescope for normal viewing, there is quite a range of settings that will give an image that appears equally sharp; you can put the image at any virtual (optical) distance from about 2 feet (60 cm) to infinity. The hand-held telescope ensures that you are placing the virtual image at infinity, so that a camera focused at infinity will be focused on it. (If you are over 50 years old and have perfect distance vision, you may be able to do without the hand-held telescope; the focusing range of your eye will be much less than a younger person's, and there is correspondingly less possible error. It's worth a try, at least.)

The effective focal length of an afocal setup is simply the focal length of the camera times the magnification of the telescope:

$$F = \begin{array}{c}\text{focal length of}\\ \text{camera lens}\end{array} \times \begin{array}{c}\text{magnification}\\ \text{of telescope}\end{array}$$

The f-ratio of the setup is the effective focal length divided by the diameter of the telescope objective:

$$f = F/\text{diameter}$$

Fig. 3.8. The moon photographed through a 12.5-cm (5-inch) Schmidt–Cassegrain telescope used as a 1250-mm f/10 telephoto lens. Half-second exposure on Kodak Technical Pan Film developed for low contrast in Technidol LC; clock drive running. (Michael Covington)

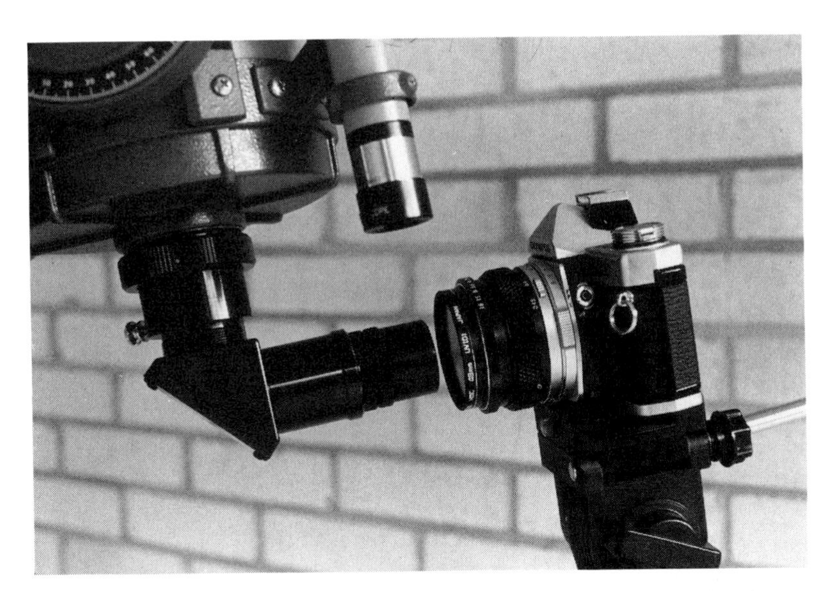

Fig. 3.9. With the afocal method, the camera and telescope can stand on separate tripods to minimize vibration.

Throughout, we use capital *F* to stand for focal length and small *f* to stand for *f*-ratio. Be sure not to confuse the two.

To get good pictures of the whole of the sun's or moon's disk, you will probably need to use your camera with its normal lens and a telescope with a magnification of 20 to 50. (The magnification, in turn, equals the telescope focal length divided by the focal length of the eyepiece.) For example, one of the setups we've used in the past involves a 6-inch (15-cm) diameter Newtonian telescope of 48 inches (1,220 mm) focal length, with a 32-mm eyepiece. This gives a magnification of 38 (that is, 1,220 divided by 32), which in combination with a 50-mm camera lens works out to an effective focal length of 1,900 mm (50 mm × 38) and an *f*-ratio of *f*/12.7 (1,900 mm/150 mm). The result is a ⅔-inch (17-millimeter) image of the moon on the film, enlargeable (in theory at least) to as much as 10 inches (25 cm) across. As Fig. 3.10 shows, excellent resolution can be achieved.

The afocal method also works well with smaller telescopes, spotting scopes, and binoculars. If you put a camera with a 50-mm lens behind one of the eyepieces of a pair of 7 × 35 binoculars, you get an effective focal length of 350 mm at *f*/10 – and you can use the other eyepiece for aiming. The sun or moon image with such a setup is a bit small for photographing lunar detail, but fine for eclipses, as well as for taking pictures of birds, distant scenery, and the like. There are commercially available brackets for coupling cameras to binoculars, or you can make your own.

Focusing

Cameras coupled to telescopes are notoriously hard to focus, but precise focusing is essential if the picture is to be sharp. To focus accurately, concentrate on low-contrast detail that is visible only when the focus is perfect, and wait for moments of exceptionally steady air.

Microprism or split-image focusing screens do not work at high *f*-ratios. Instead, focus on a plain matte screen (or a plain matte area of the existing screen if it is not interchangeable or if you choose not to change it). Extra-bright focusing screens are useful, but clear-crosshairs focusing screens (as used in photomicrography) are not; they tend to make the image look in focus even when it isn't.

You may want to focus afresh from time to time, not only to keep from accumulating errors, but also because thermal expansion or contraction of the telescope tube can cause a gradual shift.

Fig. 3.10. *Full moon. A 15-cm (6-inch) f/8 telescope with 32-mm eyepiece coupled afocally to camera with 45-mm lens, 1/125 second on Tri-X developed 7 minutes at 22°C (72°F) in D-19 diluted 1:1, printed on Kodak Polymax RC paper with #4 filter. In this case the moon is about to go into eclipse, as evidenced by a slight darkening at the lower right. (Michael Covington)*

CHAPTER FOUR

THE ANNULAR SOLAR ECLIPSE OF MAY 10, 1994, IN NORTH AMERICA

On May 10, 1994, much of the United States, as well as parts of Mexico and Canada, will see an annular eclipse of the sun. At maximum, 94% of the sun will be covered. Annularity passes through or near the cities of El Paso, Amarillo, Wichita, Chicago, Indianapolis, Detroit, Toronto, Buffalo, Saratoga Springs (NY), Burlington (Vermont), and Halifax (Nova Scotia). Off to the side of the zone from which annularity can be seen, throughout North America, people will see a partial eclipse. Annularity will last about 5½ to 6 minutes in the United States. The partial phases, whether you are inside or outside the zone of annularity, will last hours.

Site selection

To see the moon centered in front of the sun, you'll need to be very close to the center of the path.

The weather outlook is best on the coast of Sonora in Mexico, where clear skies are almost a certainty. There, however, the eclipse is slightly shorter than elsewhere (5 minutes 20 seconds in Sonora compared to 6 minutes 8 seconds in Detroit), and the sun is lower in the sky (about 40° from the horizon).

In the United States, the weather outlook is best in New Mexico and the deserts of southwest Texas. The region from central Texas to Detroit is distinctly worse, and the region east of Detroit, slightly worse yet. Unfortunately, May is one of the rainiest months of the year at these sites.

Time zones

The zone of annularity passes through the Mountain, Central, Eastern, and Atlantic time zones. Daylight Saving Time (summer time) will be in effect in Canada and most of the United States, but not Indiana, Arizona, or Mexico. Local times are shown on Maps 4.1–4.4, but changes are possible and all time zone information should be checked locally shortly before the eclipse.

Maps and tables

Maps 4.1–4.4 show the path of the zone of annularity, with times at 10-minute intervals. Table 4.1 gives the same information in numerical form,

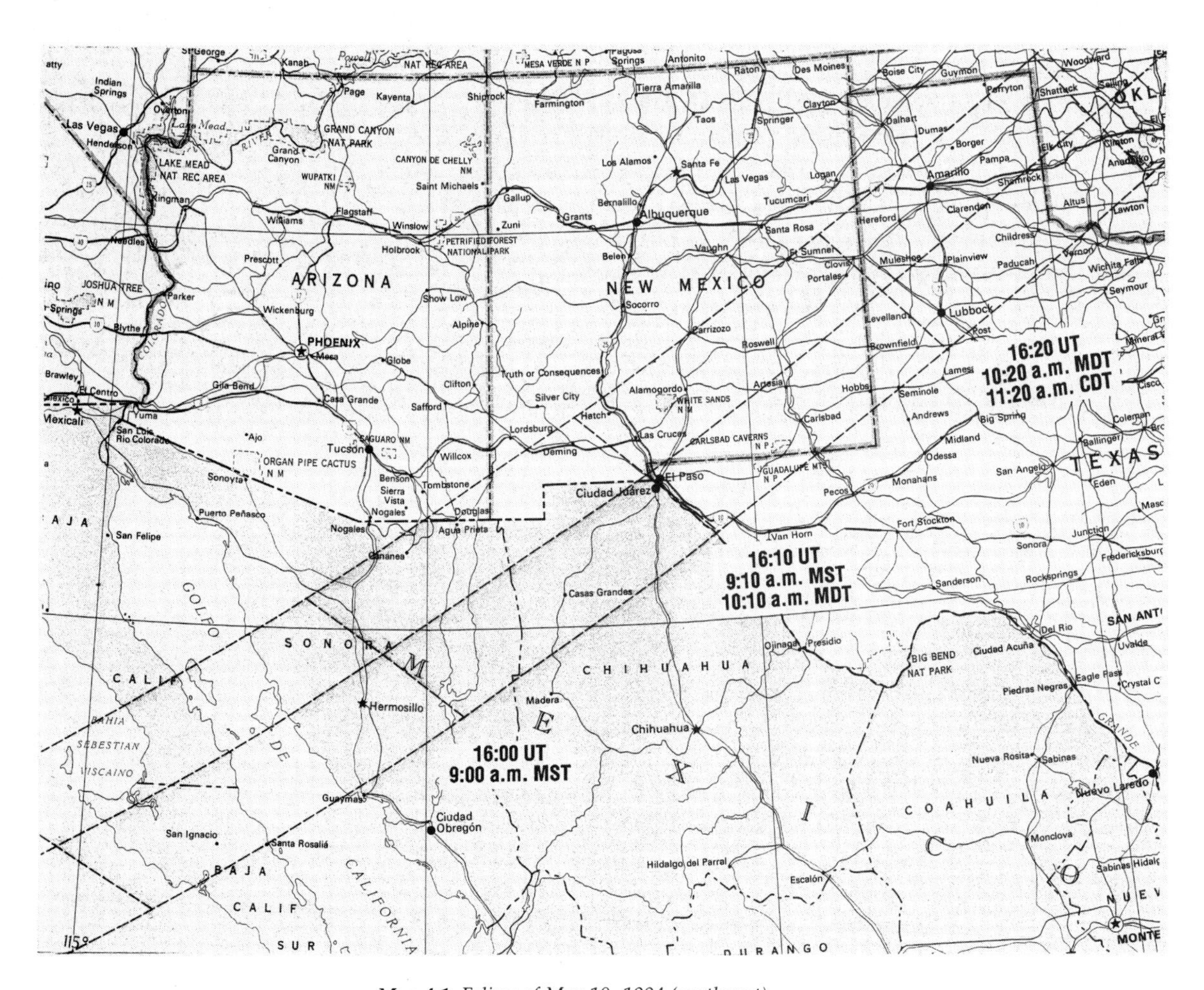

Map 4.1. Eclipse of May 10, 1994 (southwest).

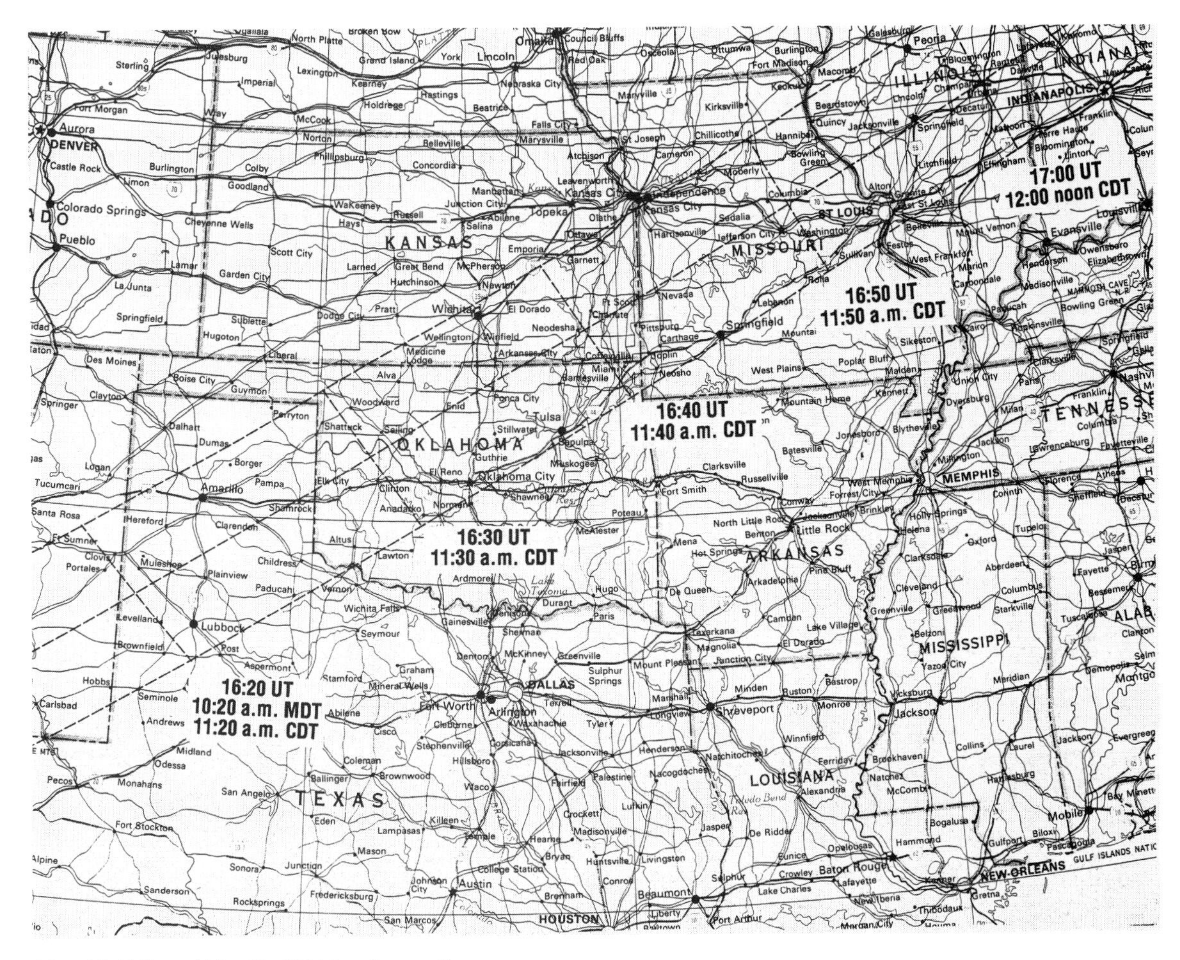

Map 4.2. Eclipse of May 10, 1994 (south central).

expressing latitude and longitude as functions of time. Table 4.2, interpolated from Table 4.1, gives latitude as a function of longitude for easier plotting on maps. Table 5.1 (p. 87) provides additional information.

Annular and partial phases

Since at most 94% of the sun will be covered, the corona will never be visible and the sky will always be blue. Solar filters or a projected image must be used at all times, whether you are in or out of the band of annularity. Indeed those seeing only partial phases will not be much worse off than those in the band of annularity.

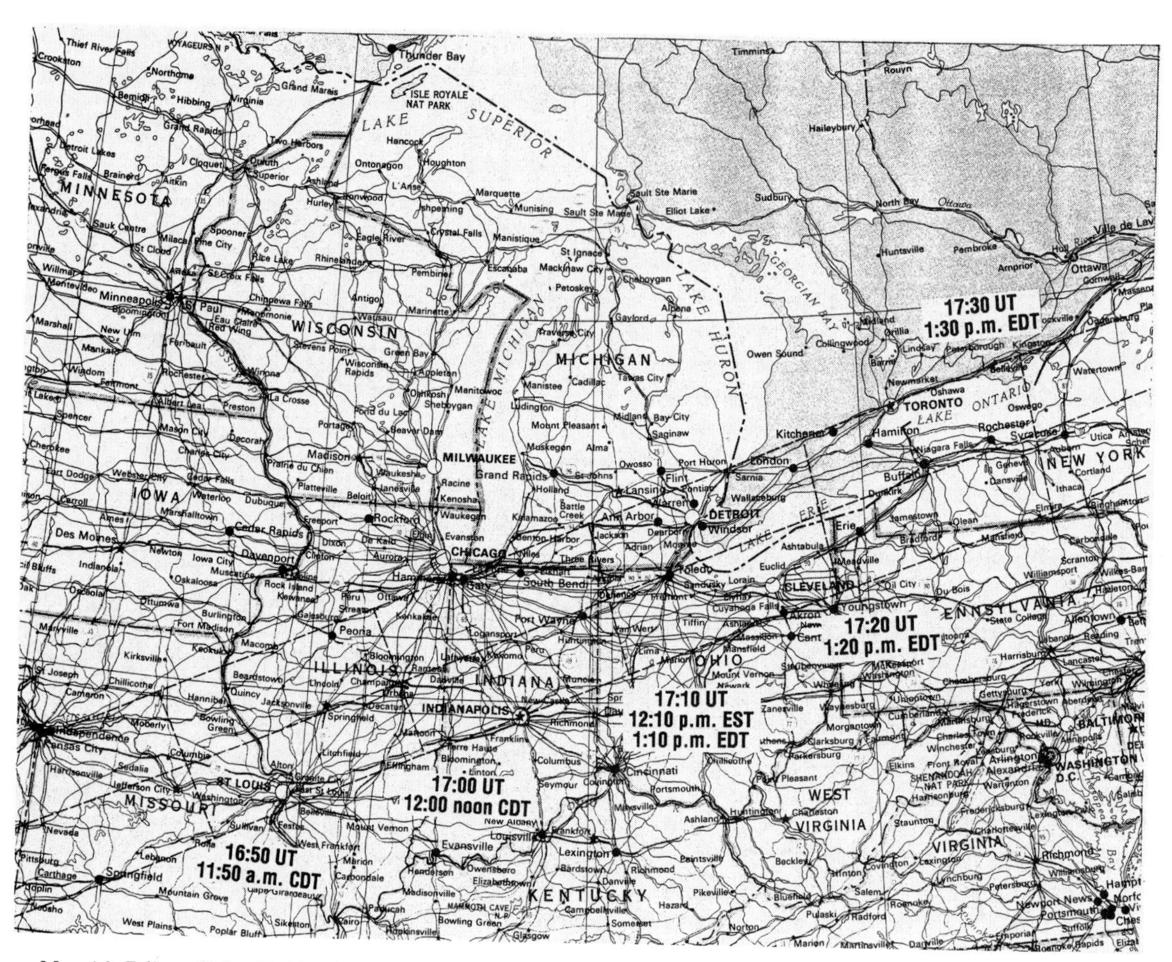

Map 4.3. *Eclipse of May 10, 1994 (north central).*

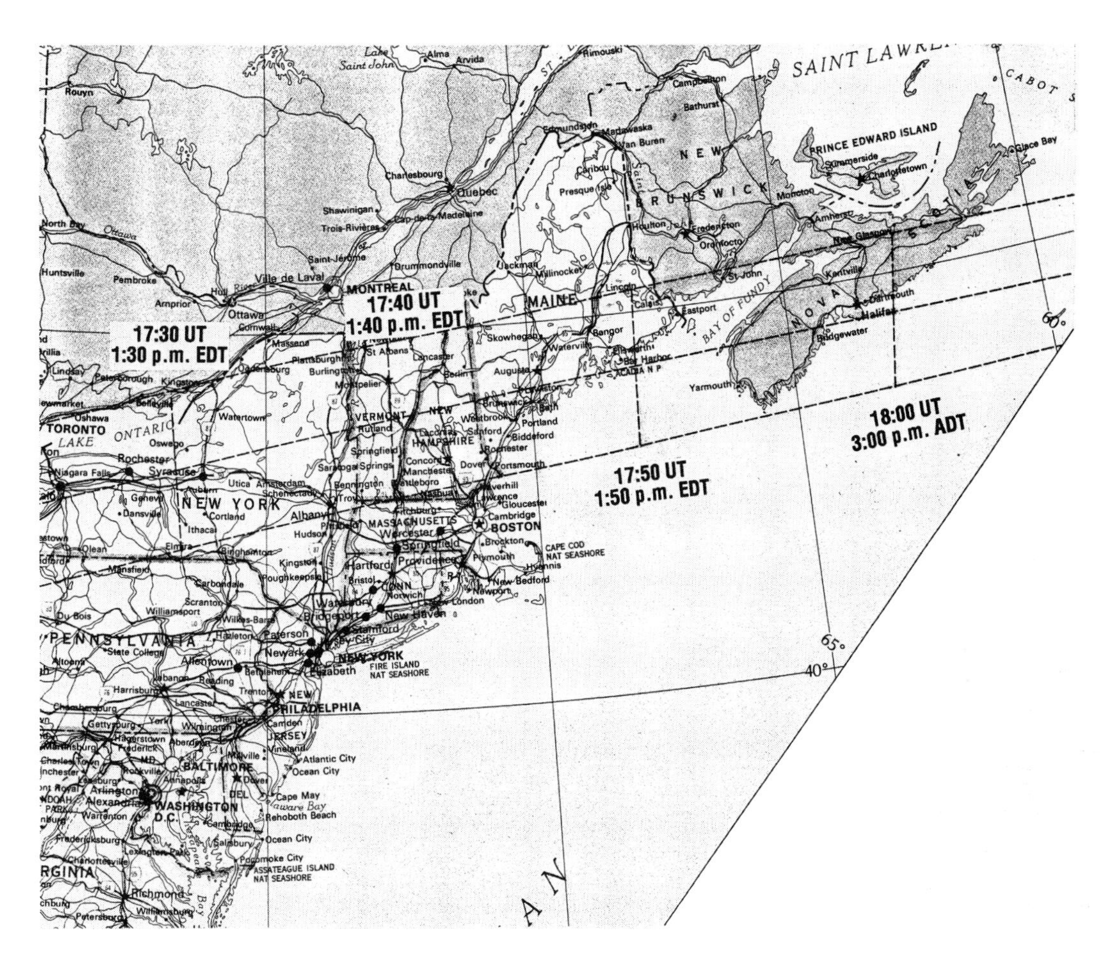

Map 4.4. Eclipse of May 10, 1994 (northeast).

Table 4.1. *Path of annularity, Tuesday, May 10, 1994. From* Circular No. 170, *US Naval Observatory.*

UT		Northern limit				Central line				Southern limit				Central line		
		Latitude		Longitude		Latitude		Longitude		Latitude		Longitude		Duration		Alt.
h	m	°	′	°	′	°	′	°	′	°	′	°	′	m	s	°
LIMIT		+14	47.0	−146	54.1	+13	33.5	−146	07.7	+12	20.9	−145	22.6	4	31.0	
15	25	+15	60.0	−143	12.2	+16	38.2	−137	25.9	+16	36.1	−133	54.1	4	41.0	9
15	30	+20	17.9	−132	03.8	+19	54.3	−129	37.7	+19	27.2	−127	27.7	4	52.2	18
15	35	+22	41.4	−126	45.8	+22	08.5	−124	52.0	+21	34.3	−123	06.5	5	00.3	24
15	40	+24	36.5	−122	49.2	+23	59.2	−121	11.3	+23	21.3	−119	38.9	5	07.1	29
15	45	+26	16.2	−119	34.0	+25	36.1	−118	06.0	+24	55.8	−116	42.4	5	13.2	33
15	50	+27	45.8	−116	44.1	+27	03.7	−115	23.3	+26	21.6	−114	06.1	5	18.9	37
15	55	+29	07.8	−114	11.6	+28	24.1	−112	56.3	+27	40.6	−111	44.2	5	24.1	40
16	00	+30	23.9	−111	51.3	+29	38.9	−110	40.6	+28	54.1	−109	32.7	5	29.1	43
16	05	+31	35.1	−109	40.1	+30	48.9	−108	33.3	+30	03.0	−107	29.1	5	33.7	46
16	10	+32	42.1	−107	35.6	+31	54.9	−106	32.4	+31	08.0	−105	31.4	5	38.1	49
16	15	+33	45.5	−105	36.3	+32	57.2	−104	36.2	+32	09.4	−103	38.2	5	42.2	51
16	20	+34	45.5	−103	40.7	+33	56.3	−102	43.6	+33	07.5	−101	48.5	5	46.1	53
16	25	+35	42.6	−101	47.9	+34	52.4	−100	53.6	+34	02.7	−100	01.2	5	49.6	55
16	30	+36	36.8	−99	56.9	+35	45.8	−99	05.4	+34	55.2	−98	15.6	5	52.9	57
16	35	+37	28.4	−98	07.1	+36	36.4	−97	18.4	+35	45.0	−96	31.2	5	55.9	59
16	40	+38	17.5	−96	17.9	+37	24.6	−95	31.9	+36	32.2	−94	47.4	5	58.7	61
16	45	+39	04.0	−94	28.7	+38	10.3	−93	45.5	+37	17.1	−93	03.6	6	01.1	62
16	50	+39	48.2	−92	39.1	+38	53.5	−91	58.7	+37	59.5	−91	19.5	6	03.2	63
16	55	+40	30.0	−90	48.5	+39	34.4	−90	11.0	+38	39.5	−89	34.7	6	04.9	64
17	00	+41	09.3	−88	56.7	+40	12.9	−88	22.2	+39	17.2	−87	48.8	6	06.4	65
17	05	+41	46.3	−87	03.2	+40	49.0	−86	31.9	+39	52.5	−86	01.5	6	07.4	66
17	10	+42	20.8	−85	07.6	+41	22.7	−84	39.6	+40	25.4	−84	12.5	6	08.1	66
17	15	+42	52.8	−83	09.7	+41	54.0	−82	45.2	+40	55.8	−82	21.4	6	08.5	66

Table 4.1. (continued)

UT		Northern limit				Central line				Southern limit				Central line		
		Latitude		Longitude		Latitude		Longitude		Latitude		Longitude		Duration		Alt.
h	m	°	′	°	′	°	′	°	′	°	′	°	′	m	s	°
17	20	+43	22.3	−81	09.1	+42	22.7	−80	48.3	+41	23.8	−80	28.0	6	08.4	65
17	25	+43	49.1	−79	05.5	+42	48.8	−78	48.5	+41	49.2	−78	31.9	6	08.0	65
17	30	+44	13.2	−76	58.5	+43	12.2	−76	45.7	+42	11.9	−76	32.9	6	07.1	64
17	35	+44	34.4	−74	47.8	+43	32.8	−74	39.3	+42	31.9	−74	30.7	6	05.9	63
17	40	+44	52.6	−72	33.1	+43	50.4	−72	29.2	+42	49.0	−72	24.9	6	04.2	62
17	45	+45	07.6	−70	14.0	+44	04.9	−70	14.9	+43	03.1	−70	15.2	6	02.2	60
17	50	+45	19.3	−67	50.0	+44	16.2	−67	56.1	+43	14.0	−68	01.2	5	59.7	58
17	55	+45	27.3	−65	20.7	+44	24.0	−65	32.3	+43	21.5	−65	42.5	5	56.9	57
18	00	+45	31.5	−62	45.6	+44	28.0	−63	02.9	+43	25.3	−63	18.6	5	53.6	54
18	05	+45	31.5	−60	03.9	+44	28.0	−60	27.4	+43	25.3	−60	48.9	5	50.0	52
18	10	+45	27.0	−57	14.9	+44	23.6	−57	45.0	+43	21.0	−58	12.7	5	46.0	50
18	15	+45	17.4	−54	17.5	+44	14.4	−54	54.7	+43	12.1	−55	29.0	5	41.6	47
18	20	+45	02.2	−51	10.4	+43	59.8	−51	55.3	+42	58.0	−52	36.7	5	36.8	45
18	25	+44	40.6	−47	51.6	+43	39.1	−48	45.0	+42	38.1	−49	34.3	5	31.7	42
18	30	+44	11.5	−44	18.5	+43	11.3	−45	21.4	+42	11.5	−46	19.5	5	26.1	39
18	35	+43	33.4	−40	27.0	+42	35.1	−41	41.0	+41	36.9	−42	49.1	5	20.0	35
18	40	+42	43.9	−36	10.6	+41	48.3	−37	38.2	+40	52.4	−38	58.3	5	13.4	31
18	45	+41	38.6	−31	17.4	+40	47.3	−33	03.0	+39	55.0	−34	38.9	5	06.2	27
18	50	+40	09.0	−25	22.0	+39	25.2	−27	35.8	+38	39.0	−29	34.9	4	58.0	22
18	55	+37	47.3	−17	02.3	+37	23.4	−20	22.3	+36	50.9	−23	07.6	4	47.9	15
19	00	...	...	....	...	...	...	....	...	+33	23.1	−11	58.9	.	...	
LIMIT		+33	28.1	−3	14.4	+32	16.6	−4	08.8	+31	05.8	−5	01.2	4	29.2	

Table 4.2. *Path of annularity, Tuesday, May 10, 1994. Interpolated table of latitude against longitude, for plotting on your own maps.*

Longitude	Latitude			Longitude	Latitude		
	Northern limit	Center of path	Southern limit		Northern limit	Center of path	Southern limit
W 120.0	N 26.045	24.601	23.170	W 90.0	N 40.790	39.640	38.501
W 119.0	N 26.566	25.125	23.698	W 89.0	N 41.136	39.997	38.869
W 118.0	N 27.092	25.655	24.232	W 88.0	N 41.469	40.340	39.222
W 117.0	N 27.622	26.190	24.771	W 87.0	N 41.788	40.668	39.560
W 116.0	N 28.157	26.730	25.315	W 86.0	N 42.093	40.981	39.883
W 115.0	N 28.694	27.273	25.864	W 85.0	N 42.383	41.280	40.191
W 114.0	N 29.235	27.819	26.416	W 84.0	N 42.659	41.565	40.483
W 113.0	N 29.777	28.368	26.972	W 83.0	N 42.922	41.835	40.760
W 112.0	N 30.320	28.919	27.530	W 82.0	N 43.171	42.091	41.022
W 111.0	N 30.863	29.470	28.088	W 81.0	N 43.407	42.333	41.270
W 110.0	N 31.405	30.021	28.647	W 80.0	N 43.628	42.560	41.504
W 109.0	N 31.946	30.571	29.206	W 79.0	N 43.837	42.774	41.723
W 108.0	N 32.484	31.119	29.764	W 78.0	N 44.033	42.974	41.927
W 107.0	N 33.019	31.665	30.319	W 77.0	N 44.216	43.161	42.117
W 106.0	N 33.550	32.207	30.872	W 76.0	N 44.386	43.334	42.294
W 105.0	N 34.075	32.742	31.419	W 75.0	N 44.543	43.494	42.457
W 104.0	N 34.593	33.273	31.961	W 74.0	N 44.688	43.642	42.607
W 103.0	N 35.105	33.796	32.497	W 73.0	N 44.821	43.776	42.743
W 102.0	N 35.609	34.312	33.025	W 72.0	N 44.942	43.898	42.866
W 101.0	N 36.104	34.820	33.544	W 71.0	N 45.051	44.007	42.977
W 100.0	N 36.589	35.318	34.055	W 70.0	N 45.148	44.105	43.075
W 99.0	N 37.063	35.807	34.556	W 69.0	N 45.235	44.191	43.161
W 98.0	N 37.528	36.283	35.046	W 68.0	N 45.310	44.266	43.235
W 97.0	N 37.981	36.748	35.525	W 67.0	N 45.374	44.329	43.296
W 96.0	N 38.422	37.202	35.990	W 66.0	N 45.426	44.380	43.346
W 95.0	N 38.849	37.643	36.443	W 65.0	N 45.468	44.420	43.384
W 94.0	N 39.264	38.070	36.883	W 64.0	N 45.500	44.449	43.410
W 93.0	N 39.666	38.484	37.310	W 63.0	N 45.521	44.467	43.425
W 92.0	N 40.055	38.883	37.722	W 62.0	N 45.532	44.475	43.430
W 91.0	N 40.430	39.269	38.119	W 61.0	N 45.533	44.472	43.424

CHAPTER FIVE

SOLAR ECLIPSES THROUGH 1999

Since total solar eclipses occur about every eighteen months, several can be observed in the next few years. However, most will occur in parts of the globe distant from most readers of this book.

The next eclipses are:

November 3, 1994	total	South America
October 24, 1995	total	Southeast Asia
March 9, 1997	total	Northeast Asia
February 26, 1998	total	South and Central America, Caribbean
August 11, 1999	total	Europe

Let us consider each of these in turn. Tracks are shown in Fig. 5.1. We will also briefly consider, for completeness, the annular eclipses that occur during the period:

April 29, 1995	annular	South America
August 22, 1998	annular	Indonesia, Malaysia
February 16, 1999	annular	Australia

It is fun to view a series of total solar eclipses, since the shape of the corona changes with the sunspot cycle (Fig. 5.2). At solar sunspot minimum, we see mainly equatorial streamers, and the corona is oblong. At solar maximum – for the 1999 eclipse, for example – there is so much activity all over that the corona appears round.

November 3, 1994, total solar eclipse

The November 3, 1994, total eclipse has favorable astronomical circumstances. The eclipse will cross South America (Fig. 5.3) from the confluence of Peru, Chile, and Bolivia to Brazil. Totality will last about 2 minutes 45

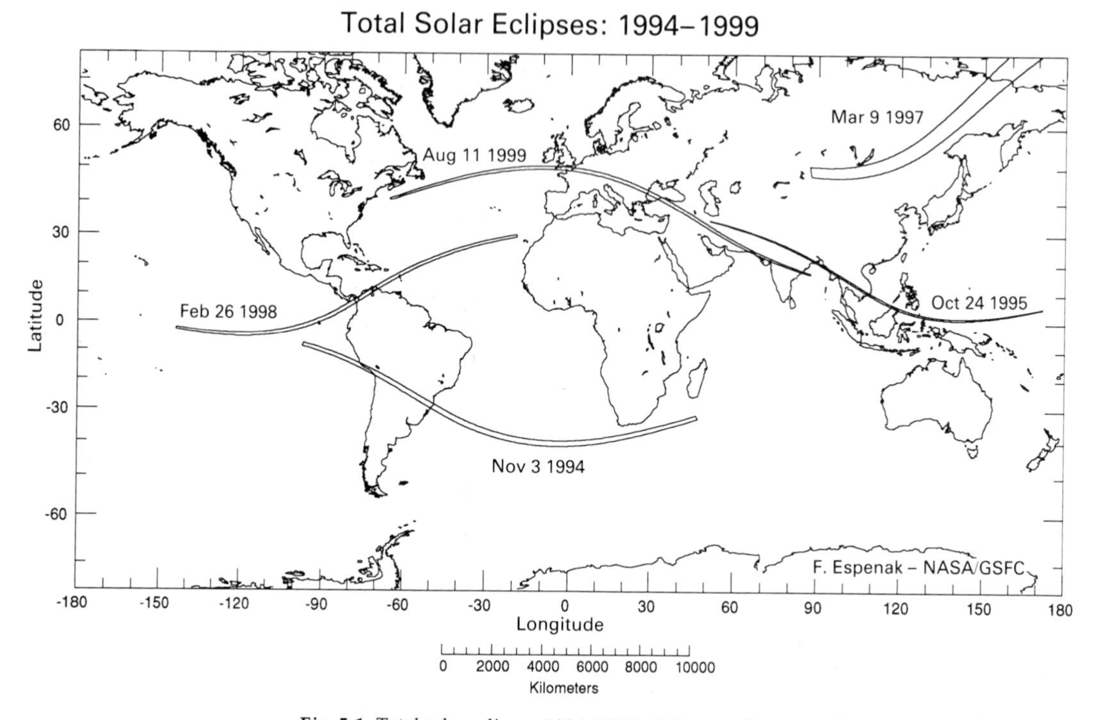

Fig. 5.1. Total solar eclipses: 1994–1999. Eclipse predictions and map courtesy of F. Espenak – NASA/GSFC.

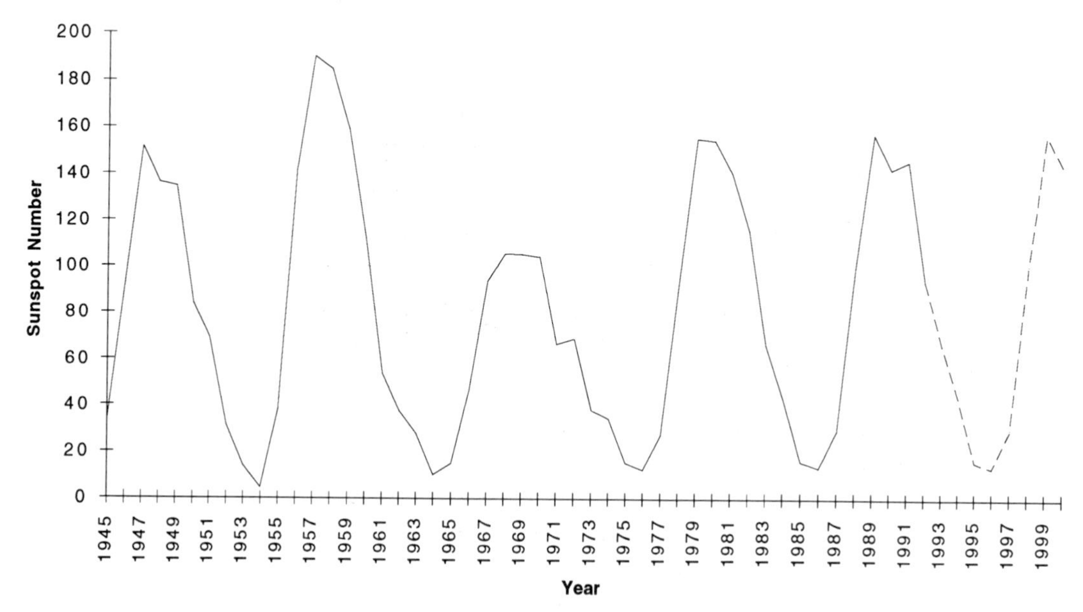

Fig. 5.2. Sunspot number, plotted as a function of time. Data are extrapolated into the future, assuming the present cycle is like the last one, to show that we will go through sunspot minimum in the mid-1990s and then return to sunspot maximum at the turn of the century. Sunspot number is defined as 10 times the number of sunspot groups on the face of the sun at any one time plus the number of individual sunspots; it is the traditional measure of solar activity.

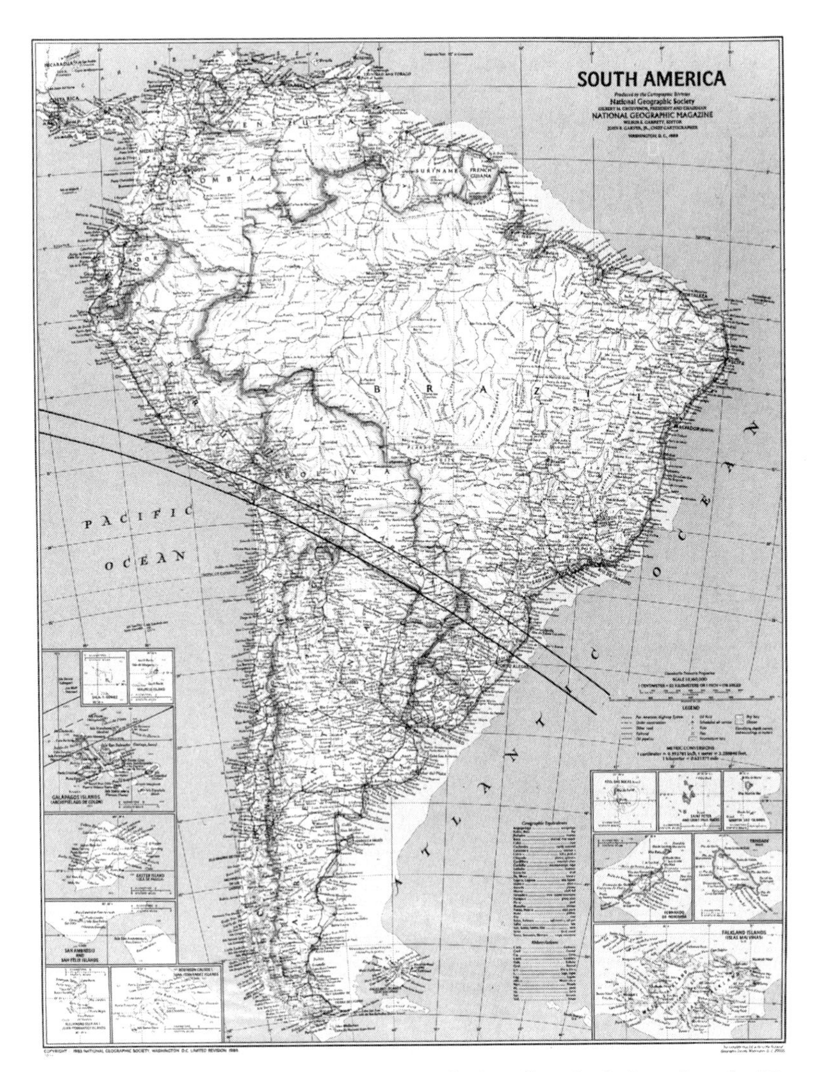

Fig. 5.3. The path of the November 3, 1994, total solar eclipse. In the inset the path of the 1998 eclipse over the Galápagos Islands is shown. (background map © National Geographic Society; eclipse plotted by Kevin Reardon and Jay M. Pasachoff)

seconds in the western part, with the sun at an altitude of about 25° above the horizon, as it rises in the early morning. Totality will lengthen to almost 4 minutes as it leaves South America in the east, with the sun at an altitude of 55°, later in the morning. The eclipse is still lengthening, and will reach its maximum totality of 4 minutes 23 seconds in mid-Atlantic ocean.

Weather forecasts are favorable overall. The western part of the path, as soon as one leaves the coast, is in very high, dry desert, and the chance of clear weather may be about 90%. In the eastern part of the path, the chance of clear weather may be 50% though logistics may be somewhat easier.

Details of the path appear in Fig. 5.4. The center line of totality first proceeds down the coast of Peru; the northern limit is near the high, easily accessible town of Arequipa. It is most desirable to stay in such high regions while getting closer to the center line; the direct path to the center line from Arequipa takes you down to sea level. Arequipa has a long, distinguished astronomical history; Harvard had an observatory there for many years, which was succeeded by a satellite-tracking station run by the Smithsonian Astrophysical Observatory. These efforts, however, are no longer in effect.

Unfortunately, as of early 1993, the political situation in Peru is exceedingly worrisome. The control of large parts of the countryside by the Shining Path will no doubt not end with the 1992 arrest of their leader, Abimael Guzmán, and many visitors will not feel safe – indeed, may not be safe. At the time of writing, the United States government warns tourists not to go to Peru. Obviously, the situation may change before the eclipse, and we encourage you to look into the latest information.

The optimal observation locations may, thus, be in extreme northern Chile, a thousand miles north of Santiago. One must nevertheless, in most cases, fly into Santiago from the United States before flying north to the site. Towns like Sabaya may be the best. Hotels are sparse, and none are of the highest quality. Travel agents are already reserving all the available spaces, and you may be essentially forced to buy a package trip from one of them, unless you are inclined to camp or to take your chances. The advantages of the Chilean sites include high, dry desert air and a very low chance of clouds. Transport facilities should also be reasonably reliable, and self-drive

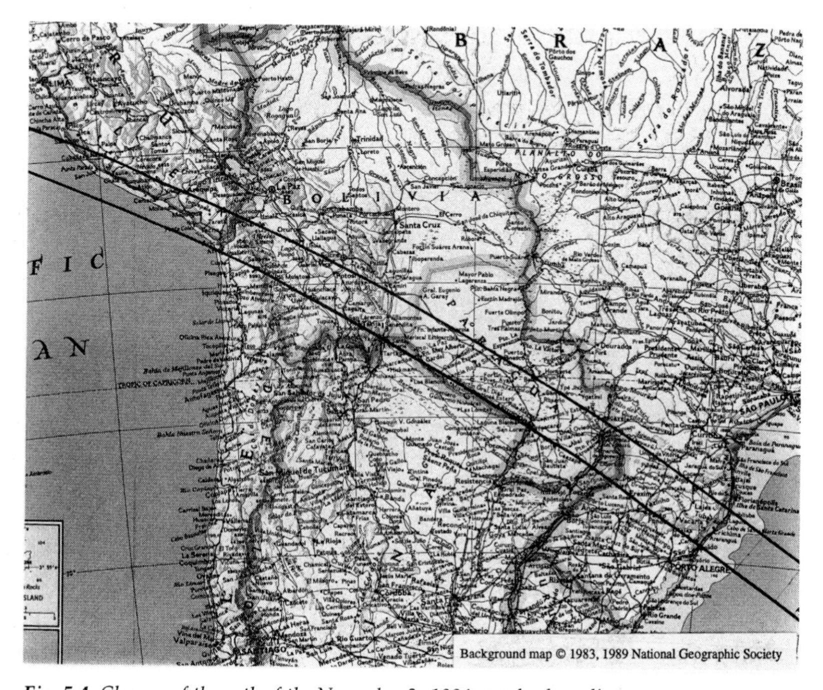

Fig. 5.4. Closeup of the path of the November 3, 1994, total solar eclipse.

rental cars can be reserved; the town of Arica is the largest in the vicinity.

The United States has a long history of astronomical work in Chile, with the US National Optical Astronomy Observatories participating in the Cerro Tololo Inter-American Observatory north of La Serena. An 8-m telescope is now planned for another peak in that vicinity, part of the Gemini project that NOAO is running with foreign participation from England, Canada, Chile, Argentina, and Brazil. The Carnegie Institution of Washington also has an observing site in that vicinity on Las Campanas; a 2.6-m telescope is there. This site is operated by the Observatories of the Carnegie Institution of Washington (and is thus the descendant of the Mt. Wilson Observatory), whose headquarters are in Pasadena, California. Some eclipse tours may include visits to astronomical sites. But all these observatories are far south of the eclipse path.

Santiago, the capital of Chile, is a cosmopolitan and European-like city, well worth a visit. Seaside resorts like Viña del Mar are also very popular. The date of the eclipse, November 3, corresponds to northern-hemisphere May's springtime.

Another set of astronomically favorable sites are in Bolivia. Sites near Lago Poopó are in similarly high, dry, relatively cloudless locations to the Chilean and Peruvian sites. Within the band of totality, the city Potosí is not far, only about 70 miles (110 km), from the large city of Sucre and 300 miles (500 km) from La Paz, the capital. I (JMP) have been advised against driving a rental car myself, so pass on the hint that it may be wise to hire a driver in La Paz if you are going to drive down. The towns of Sabaya and Villamontes are also in totality. Some tours will charter a train.

The eclipse then passes over Paraguay, whose government has undergone elections and changed considerably over the past few years. The capital, Asunción, is within totality, near its southern limit.

On the boundary shared by Paraguay, Brazil, and Argentina, the path of totality goes over Iguazú Falls, one of the wonders of the world. This set of falls is much higher (269 feet, 82 m, compared with 167 feet, 51 m, for the American Falls at Niagara) and is four times wider than Niagara. Most tourists reach it by plane from either Rio de Janeiro, Brazil, or from Buenos Aires, Argentina. My wife and I found Iguazú one of the most impressive sights we have ever seen, and certainly recommend a visit. We were impressed by both the view from the Argentinean side, which takes you to the top, and the view from the Brazilian side, where you walk below the falls. The infrastructure – roads, electricity, etc. – is excellent in the vicinity, and there are many hotels, though we did not see any of the highest caliber. Indeed, the largest dam and electricity generator in the world is nearby. The problem with this region is the weather. Surveys are now being made, but the chance of seeing the eclipse may be closer to 25:75 than the 90:10 chance in the west.

Next, the eclipse passes through southern Brazil, with the path of totality going north of Pôrto Alegre and far south of São Paulo as it reaches the coast. One could certainly routinely fly to Pôrto Alegre and drive up for the day; the centerline is about 125 miles (200 km) northeast.

Finally, after traversing the Atlantic, the path of totality will cross about 500 miles (800 km) south of Cape Town, South Africa. Out to sea, there, it will intersect the path of the June 30, 1992, total solar eclipse in a location I shall always remember fondly.

October 24, 1995, total eclipse

The October 24, 1995, eclipse passes first Iran, Afghanistan, and Pakistan, then over India (just south of Agra, Varanasi, and Calcutta) and Bangladesh. (The Taj Mahal will be, unfortunately, just outside the zone of totality.) Then, as the eclipse gets higher in the sky, it passes over the Southeast Asian countries of Myanmar (formerly Burma), Thailand, Cambodia, and Vietnam. It reaches its maximum duration of 2 minutes 10 seconds in the ocean north of Borneo (the island that is covered by parts of Malaysia and Indonesia).

As of this writing, I would certainly not consider going to Iran or Afghanistan for reasons of safety. In any case, the eclipse is only 20° high and 45 seconds or so long in Pakistan, and reaches 39° and 1 minute 15 seconds by the time it crosses India and reaches Calcutta. Next, Myanmar is another country in which I would at present not choose to visit for reasons of personal safety.

Continuing along the path, the situation in Thailand, Cambodia, and Vietnam is certainly different from what it was two and three decades ago, and I know people who travel there without problem. The United Nations helped Cambodia through elections in 1993; if the situation continues to improve, then it would be especially exciting to see the eclipse from the ancient ruins of Angkor Wat, which is in the zone of totality. It is near the city of Siem Reap. The twelfth-century temple complex is, by all accounts, fantastic to see. Hotels of international quality are present there. Totality at Angkor Wat will be almost 2 minutes long and the sun will be above 60°. But in early 1993, there were terrorist attacks in Siem Reap.

Further along the path, the centerline passes through the Vietnamese city of Ba Ra and out to the coast near Phan Thiet, about half way between Ho Chi Minh City and Cam Ranh. The trip from either would be about 100 miles (160 km), and totality will occur in the late morning.

March 9, 1997, total eclipse

The March 9, 1997, total eclipse, will be seen by very few people. Although it will last up to 2 minutes 50 seconds, and its path is as wide as 221 miles

(356 km), it passes over only relatively inaccessible regions of Mongolia and Siberia. The eclipsed sun never gets above 23° from the horizon.

The band of totality takes in the northern edge of Mongolia; its southern limit reaches about 100 miles (160 km) north of Ulan Bator, the capital (latitude 48 degrees North, about that of Seattle). The sun will be only about 13° high, though totality will be reasonably long, 2 minutes 25 seconds. The southern part of the path then clips the northernmost extension of China in that eastern part. The path continues north-northeast over Siberia, where it reaches its maximum duration and altitude, before disappearing into the Arctic.

It would be nice if a chartered aircraft were arranged to provide the clear skies of high altitude for this eclipse.

February 26, 1998, total eclipse

The February 26, 1998, eclipse will be quite the opposite in some of the observing conditions from its predecessor. It will cross the boundary of South and Central America, and then be visible from the Caribbean.

The eclipse starts in the Pacific and heads northeastward, first clipping the northern Galápagos Islands. The centerline is north of the islands. The southern limit of the path misses Isla Fernandina and crosses the northern portion of Isla Isabela. Also between the southern limit and the centerline are only the very small islands of Isla Pinta and Isla Marchena. The sites are off the centerline, where totality will be almost 4 minutes; close to the edge of the path, totality is short but Baily's beads are prolonged. Boats could reach the centerline. The larger islands of Isla Santiago (also called Isla San Salvador), Isla Santa Cruz (with its Charles Darwin Research Station), and Isla Cristóbal (where the main airport is located) and the small island Baltra (with its airport) are, unfortunately, south of the path of totality. (See the inset on the South America map, Fig. 5.3, used to plot the 1994 eclipse.)

Then the eclipse reaches its maximum duration of 4 minutes 8 seconds in the Pacific Ocean before hitting the South American continent at the boundary of Panama and Colombia. Thus all post-Galápagos viewing will be in the afternoon. The path continues along the northern part of Colombia. The northern limit is not far south of Cartagena, Colombia. On the centerline, the totality will last 4 minutes and the sun will be 70° high. The eclipse then clips northwestern Venezuela, going over Maracaibo, Venezuela.

Next the path goes over the Caribbean, first passing over two Dutch islands in the Lesser Antilles. Totality includes Aruba and almost all of Curaçao, both only about 20 miles (30 km) off the South American coast. During the 1993 eclipse-anniversary week on Aruba, there were puffy clouds each afternoon, with rain all day on one of the days. Such rain is supposedly unusual on such a dry island, but nonetheless can happen. A few weeks later, the three days observed by one of us (JMP) boasted of excellent eclipse weather. Though there were some puffy clouds, the bulk of the sky was clear at midday. Two of the mornings began cloudy but cleared up by 10 or 11 a.m. Aruba has over seven thousand hotel rooms and all the Caribbean islands in totality are readily accessible by plane from the United States.

Much farther into the Caribbean, the path of totality intersects some of the Leeward Islands. The sun is still 50° high and totality will last about 3 minutes 20 seconds. Antigua, Montserrat, and Guadeloupe are all in the path of totality. It is high tourist season there in February, and there is a reasonable chance of clear enough weather to see the eclipse. After all, one often sees the sun, which is why these islands are resorts.

The island of Antigua is part of the nation of Antigua and Barbuda, independent from Britain since 1981. The capital is St John's. The Colony of Montserrat is a British dependency, 27 miles (43 km) southwest of Antigua. The capital is Plymouth. Guadeloupe is a Département of France, considered an actual part of France rather than a colony. Both the capital, Basse-Terre, and another main island, Grande-Terre, are in the path of totality.

August 11, 1999

The 1999 total eclipse crosses over readily available locations in Europe, so we devote a separate chapter to it.

Annular eclipses

			moon:sun
April 29, 1995	annular	South America	95.0%
August 22, 1998	annular	Indonesia, Malaysia	97.3%
February 16, 1999	annular	Australia	99.3%

Tracks of these eclipses are shown in Fig. 5.5.

Since the corona (and diamond ring, etc.) are not visible at an annular eclipse, most people don't find them worth travelling to. Still, for the aficionado, they are a chance to stand in the shadow of the moon.

The April 29, 1995, annular eclipse crosses southern Ecuador (where, it turns out, Panama hats come from), northern Peru, and southern Colombia. The rest of the path crosses northern, inaccessible regions of Brazil. The eclipse lasts over 6 minutes 30 seconds in Ecuador; after all, it occurs there near the equator near noon.

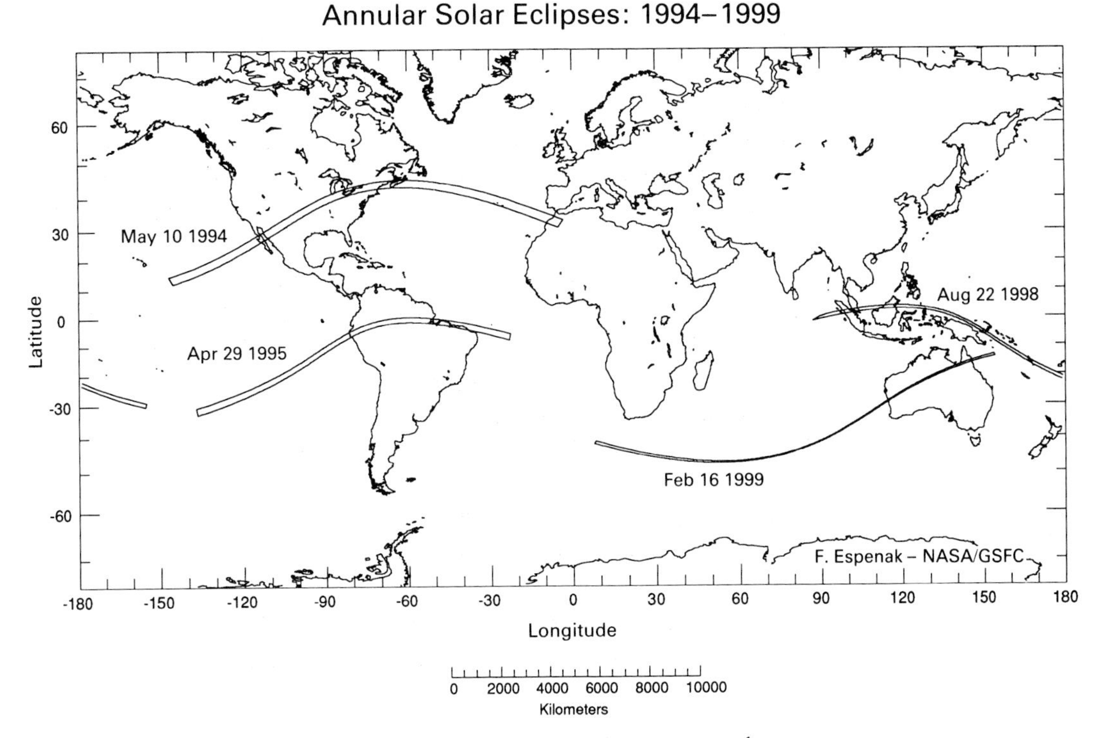

Fig. 5.5. Annular solar eclipses: 1994–1999. Eclipse predictions and map courtesy of F. Espenak – NASA/GSFC.

The eclipse will not be visible from any large cities. It is far south of Quito, Ecuador, passing over only Machala.

The August 22, 1998, annular eclipse will be visible from northern Sumatra (an island of Indonesia), where totality will be 2 minutes 50 seconds though the sun will still be low in the early morning sky, less than 15°. The sun will be about 30° high and totality slightly longer over northern Borneo (where the band of annularity passes territory of both Malaysia and Indonesia). Then the path of annularity goes out to sea. Maximum annularity will occur over the ocean north of the coast of Papua New Guinea, and good visibility could occur on the southwestern half of the island of New Britain. Totality there will be over 3 minutes 10 seconds and the sun will be 70° above the horizon.

The February 16, 1999, annular eclipse will have a fairly small rim of sun visible, though it should still be too bright to see the corona. Annularity will be visible only from northwestern Australia, in Western Australia, the Northern Territory, and northern Queensland. When the path hits the west coast of Australia, the sun will be about 45° high and totality will last about 45 seconds. No part of the path is readily accessible. Only the towns of Burketown, Karumba, and Normanton at the south end of the Bay of Carpentaria appear near the path, and they will be crossed in the late afternoon. The sun will be only about 15° high, and totality will last a few seconds over 1 minute.

Since there is no total or annular solar eclipse in the year 2000, we have now surveyed all such eclipses through the beginning of the new millennium, January 1, 2001. We have to wait until August 21, 2017, for the next total solar eclipse in the United States (Oregon to South Carolina) and April 8, 2024, for the next total solar eclipse in Canada (New Brunswick, Gaspé, Prince Edward island, and Newfoundland). Europe will have to wait for the August 12, 2026, eclipse in Spain.

Circumstances of the solar eclipses in the 1990s†

In Figs. 5.6–5.14 for each solar eclipse, an orthographic projection map of earth shows the path of penumbral (partial) eclipse. North is to the top in all cases and the daylight terminator is plotted for the instant of greatest eclipse. The sub-solar point on earth is indicated by a "*". The maps are oriented with the origin at the sub-solar longitude at greatest eclipse and at a latitude equal to the sun's declination ±45°.

The limits of the moon's penumbral shadow delineate the region of visibility of the partial solar eclipse. This irregular or saddle shaped region often covers more than half of the daylight hemisphere of earth and consists of several distinct zones or limits. At the northern and/or southern boundaries lie the limits of the penumbra's path. Partial eclipses have only one of these limits, as do central eclipses when the shadow axis falls no closer than about 0.45 radii from earth's center. Great loops at the western and eastern extremes of the penumbra's path identify the areas where the eclipse begins/ends at sunrise and sunset, respectively. The curves are connected in a distorted figure eight. Bisecting the "eclipse begins/ends at sunrise and sunset" loops is the curve of maximum eclipse at sunrise (western loop) and sunset (eastern loop). The points "P1" and "P4" mark the coordinates where the penumbral shadow first contacts (partial eclipse begins) and last contacts (partial eclipse ends) earth's surface, respectively.

A curve of maximum eclipse is the locus of all points where the eclipse is at maximum at a given time. Curves of maximum eclipse are plotted at each half hour Universal Time. They generally run from the northern to the

† This note, the maps, and the charts are by Fred Espenak, NASA/Goddard Space Flight Center.

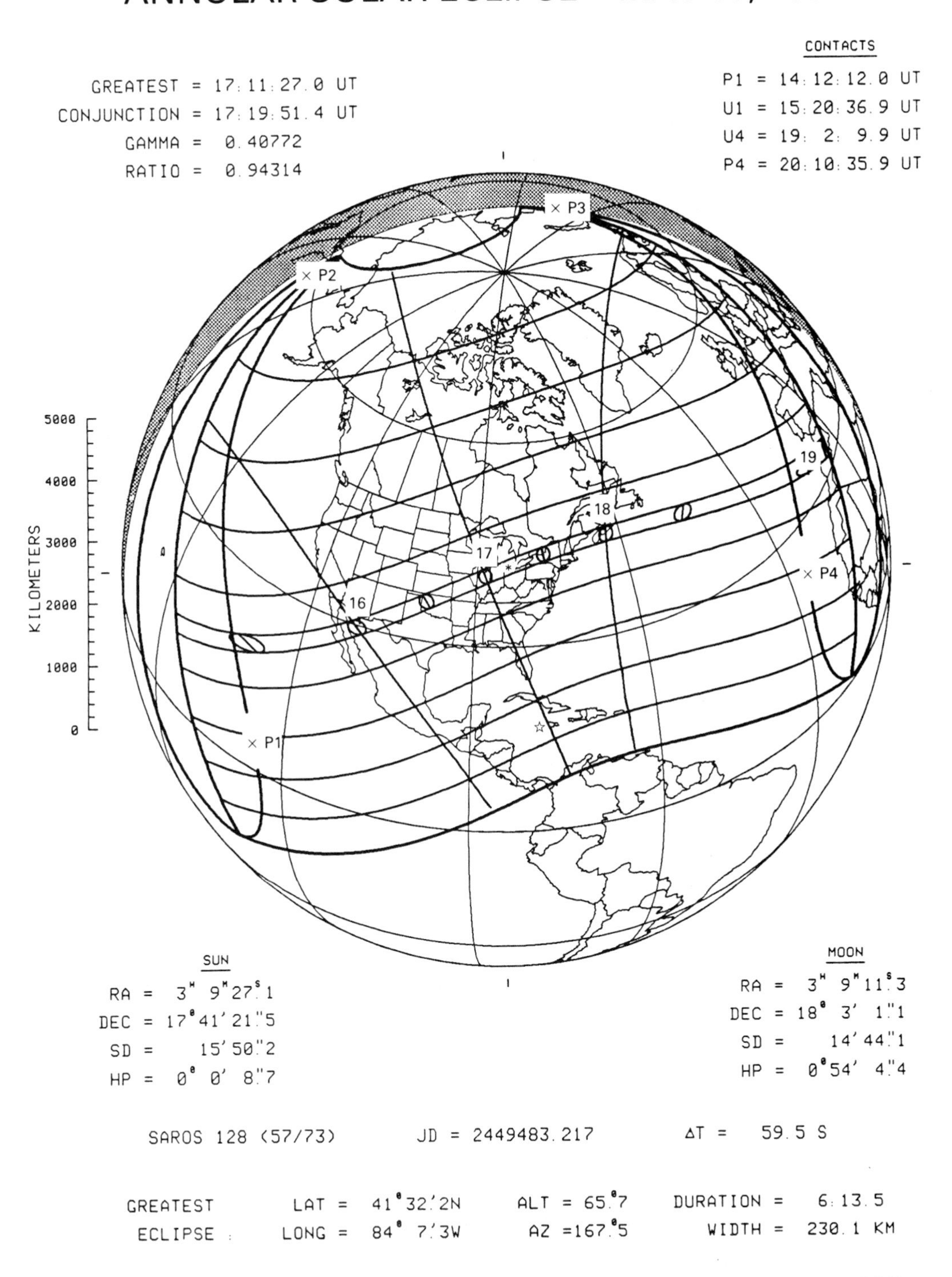

Fig. 5.6. Annular solar eclipse – May 10, 1994. Eclipse predictions and map courtesy of F. Espenak – NASA/GSFC.

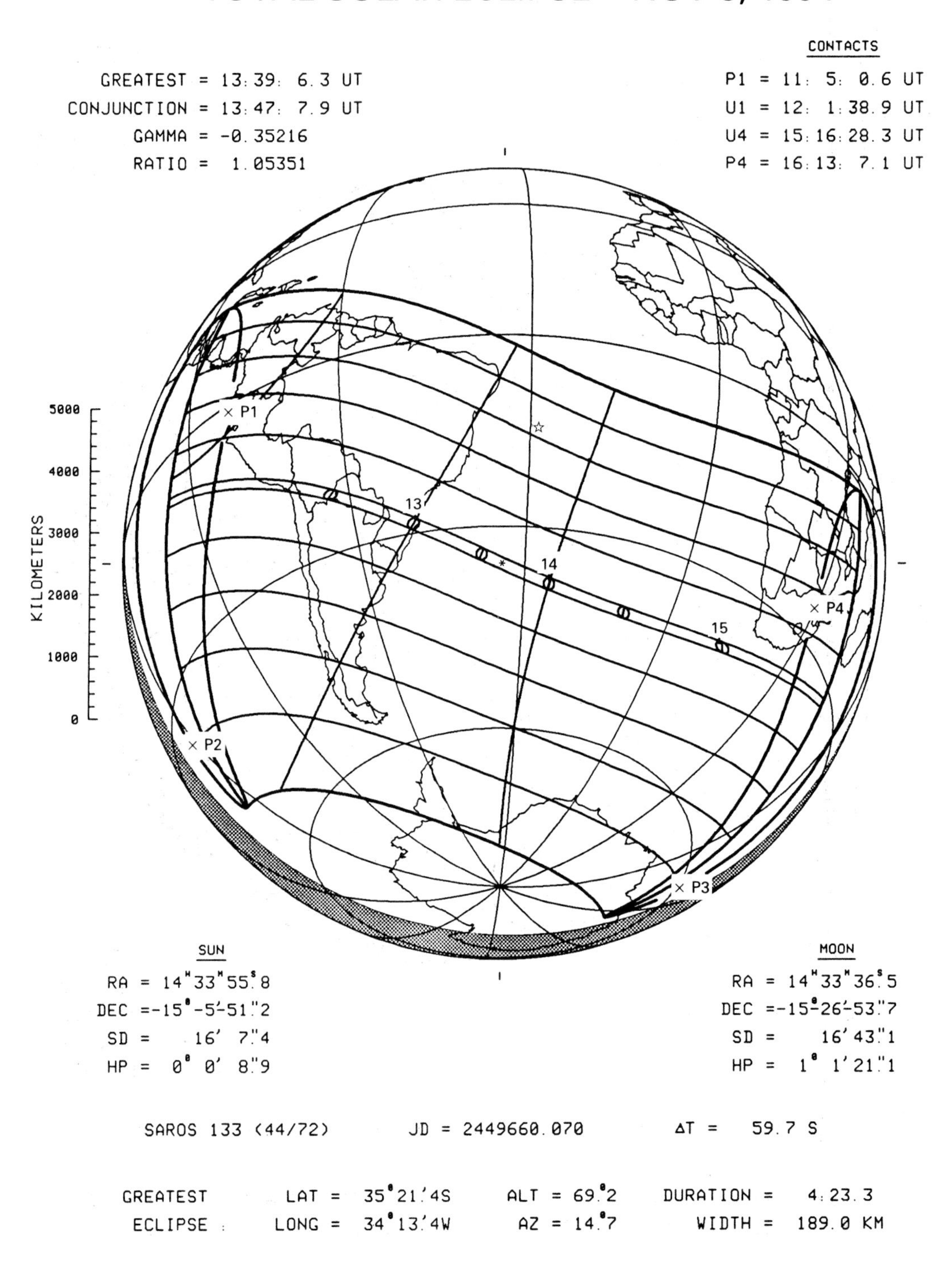

Fig. 5.7. *Total solar eclipse – November 3, 1994. Eclipse predictions and map courtesy of F. Espenak – NASA/GSFC.*

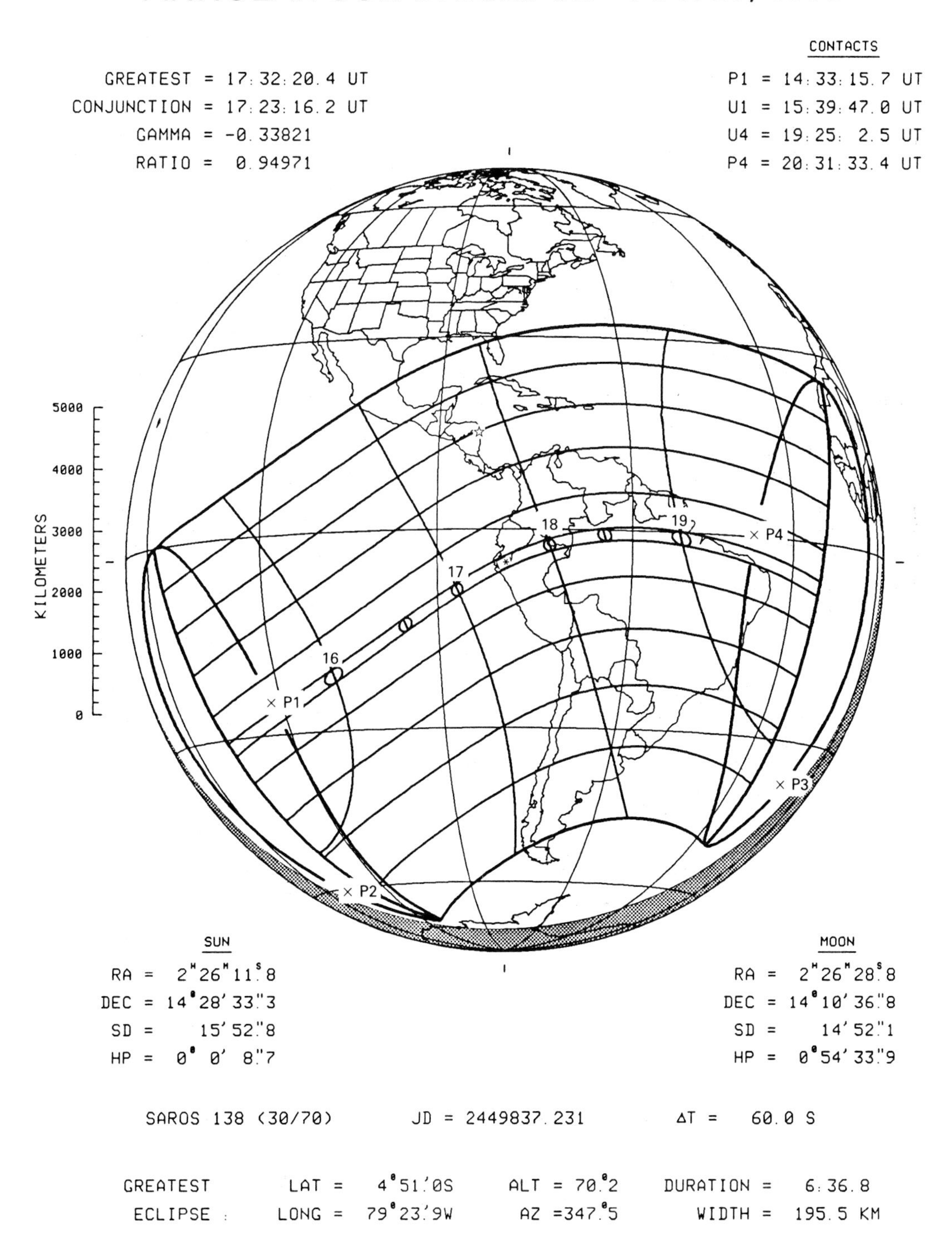

***Fig. 5.8.** Annular solar eclipse – April 29, 1995. Eclipse predictions and map courtesy of F. Espenak – NASA/GSFC.*

TOTAL SOLAR ECLIPSE – OCT 24, 1995

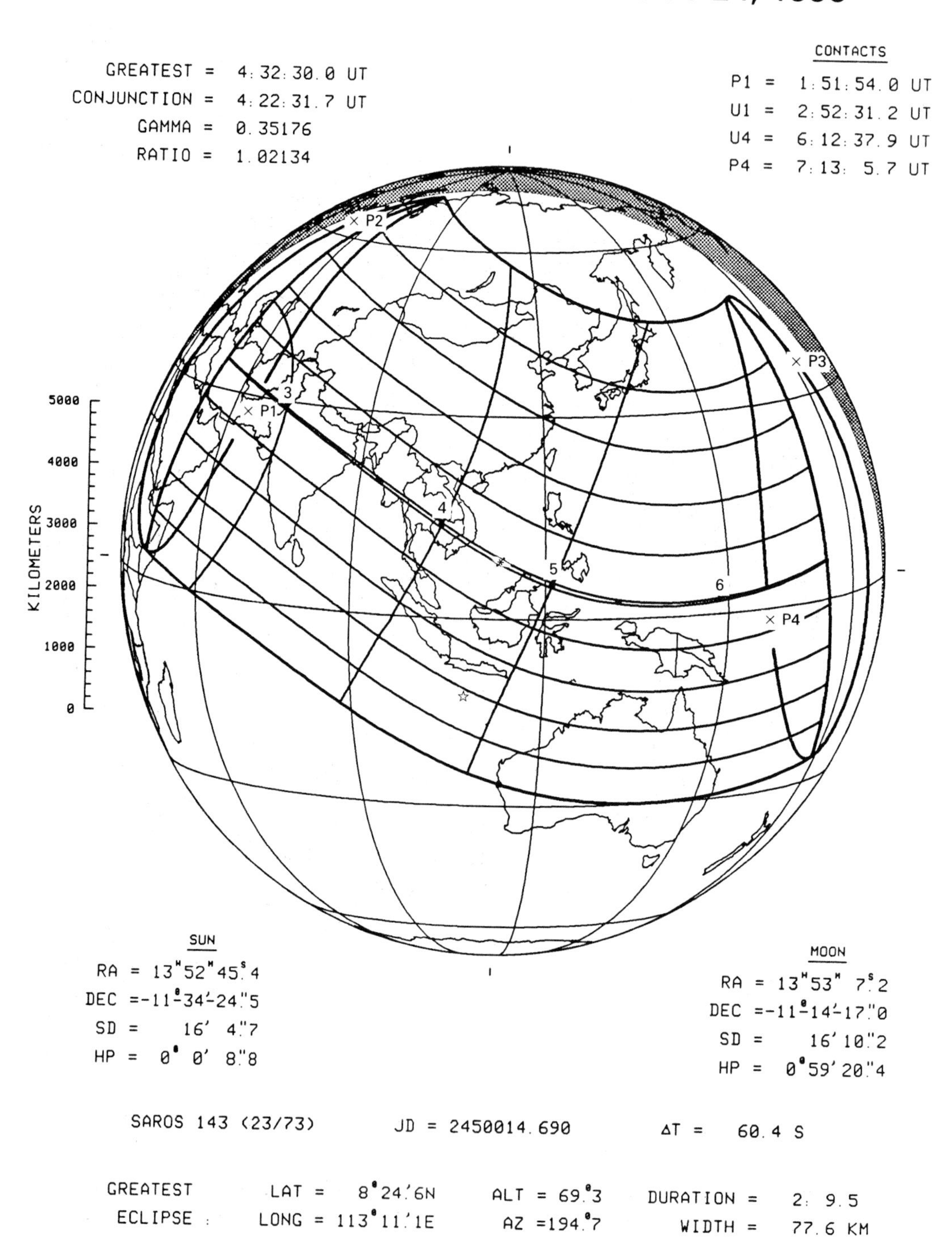

Fig. 5.9. *Total solar eclipse – October 24, 1995. Eclipse predictions and map courtesy of F. Espenak – NASA/GSFC.*

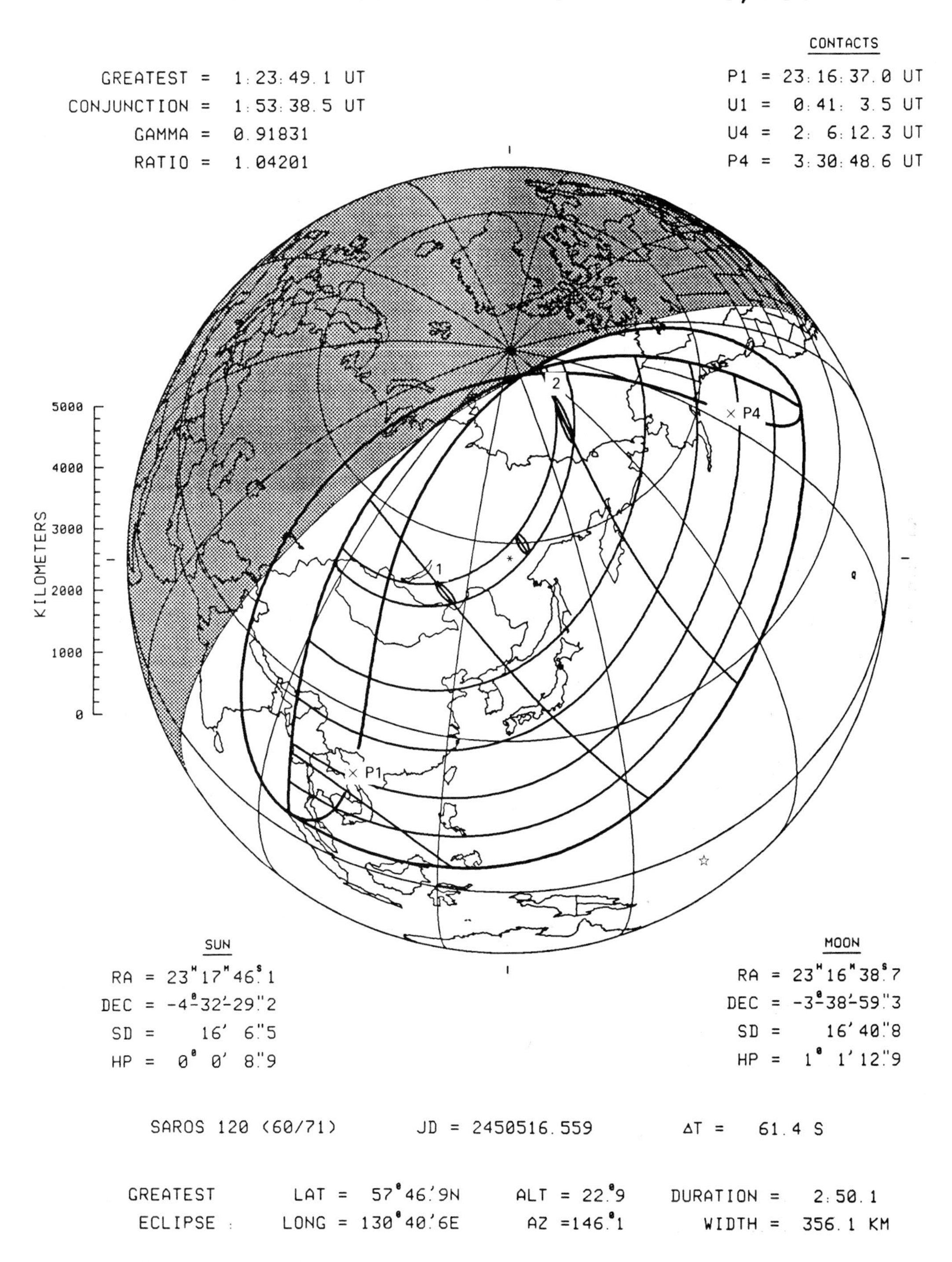

Fig. 5.10. Total solar eclipse – March 9, 1997. Eclipse predictions and map courtesy of F. Espenak – NASA/GSFC.

TOTAL SOLAR ECLIPSE – FEB 26, 1998

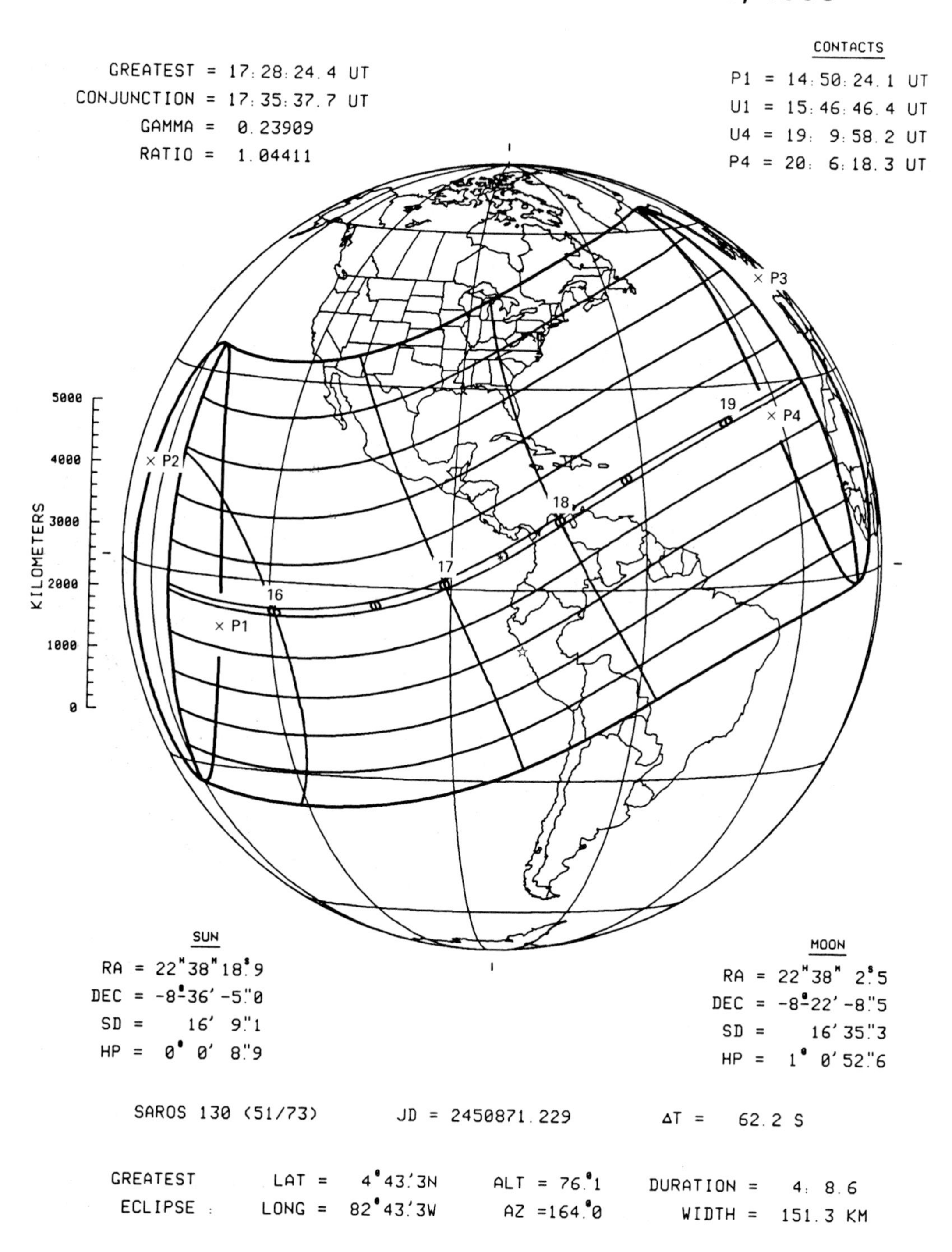

Fig. 5.11. Total solar eclipse – February 26, 1998. Eclipse predictions and map courtesy of F. Espenak – NASA/GSFC.

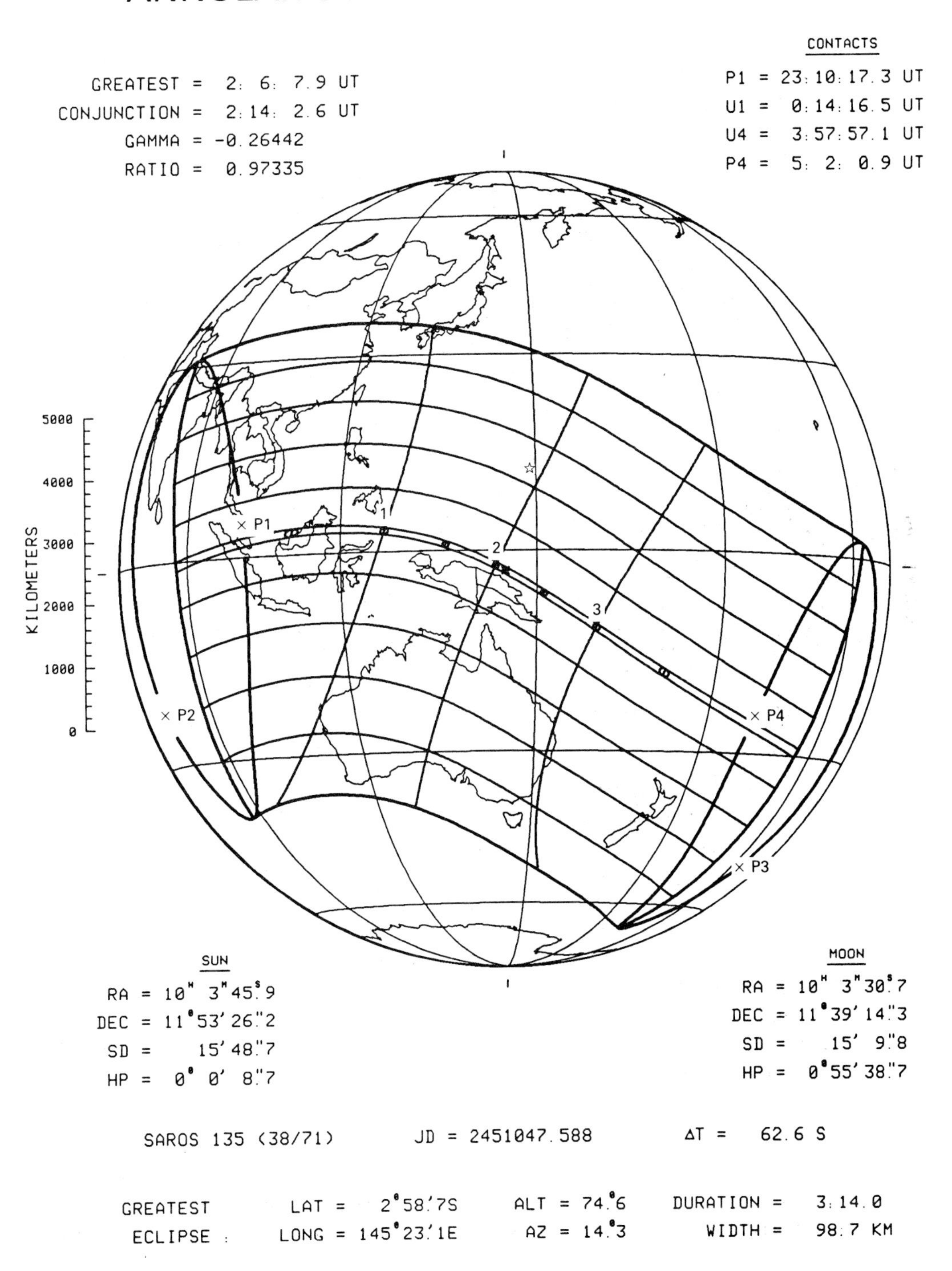

Fig. 5.12. Annular solar eclipse – August 22, 1998. Eclipse predictions and map courtesy of F. Espenak – NASA/GSFC.

ANNULAR SOLAR ECLIPSE – FEB 16, 1999

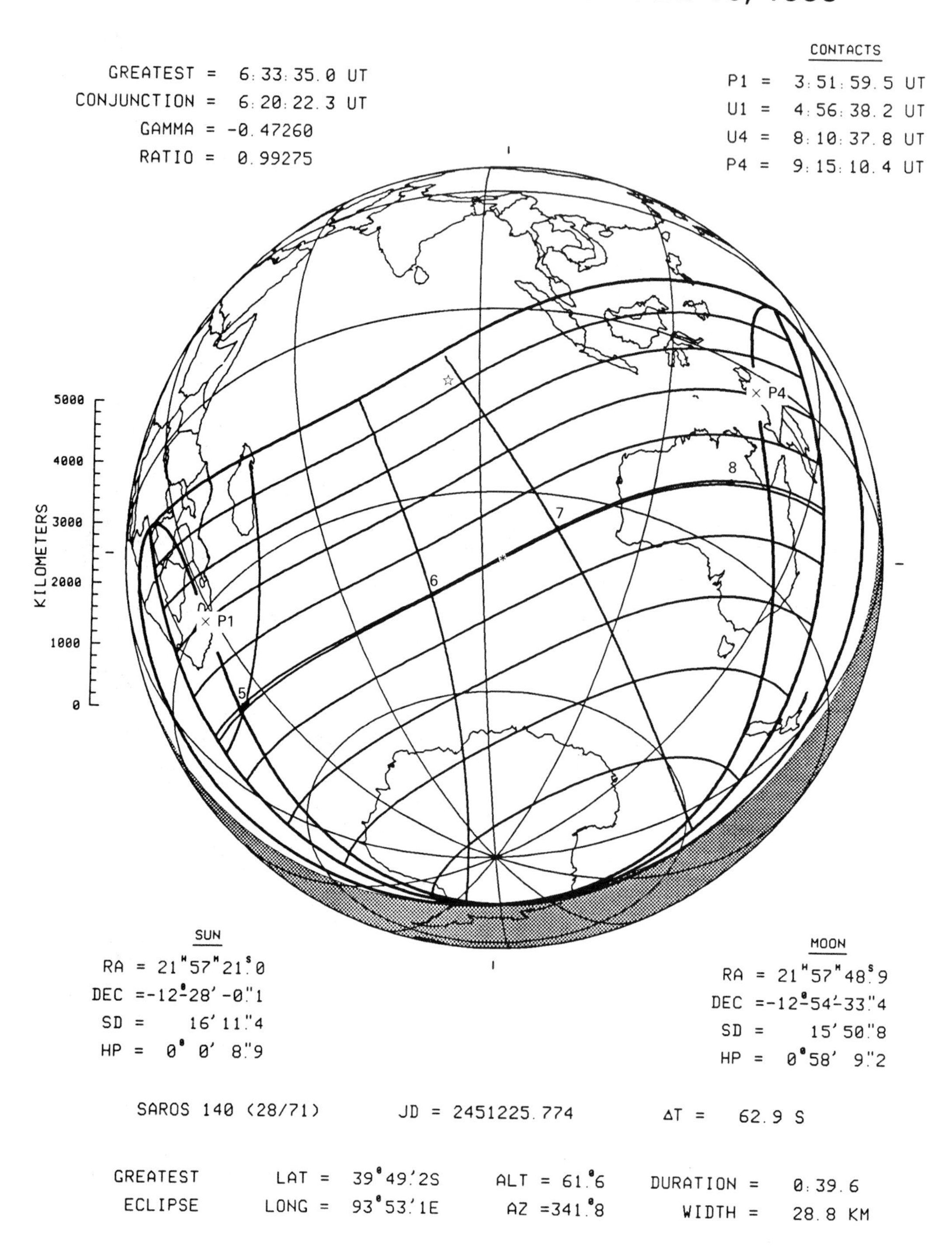

Fig. 5.13. Annular solar eclipse – February 16, 1999. Eclipse predictions and map courtesy of F. Espenak – NASA/GSFC.

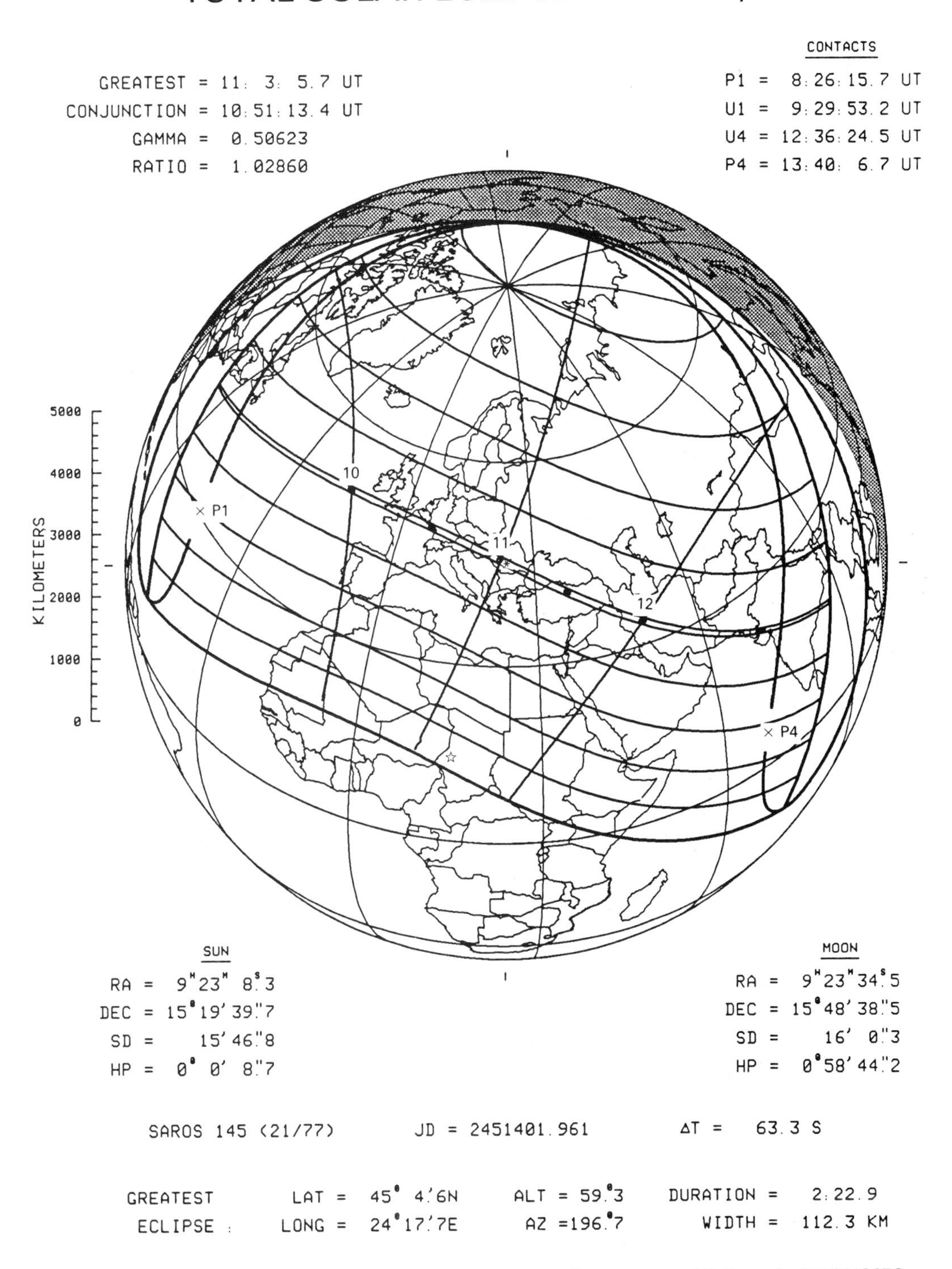

Fig. 5.14. Total solar eclipse – August 11, 1999. Eclipse predictions and map courtesy of F. Espenak – NASA/GSFC.

southern penumbral limits, or from the maximum eclipse at sunrise and sunset curves to one of the limits. The curves of constant eclipse magnitude delineate the locus of all points where the magnitude at maximum eclipse is constant. These curves run exclusively between the curves of maximum eclipse at sunrise and sunset. Furthermore, they are parallel to the northern/southern penumbral limits and the umbral paths of central eclipses. In fact, the northern and southern limits of the penumbra can be thought of as curves of constant magnitude of 0.0. The adjacent curves are for magnitudes of 0.2, 0.4, 0.6 and 0.8.

Greatest eclipse is defined as the instant when the axis of the moon's shadow passes closest to earth's center. Although greatest eclipse differs slightly from the instants of greatest magnitude and greatest duration (for total eclipses), the differences are usually negligible. The point on earth's surface nearest to the axis at greatest eclipse is marked by an asterisk. For partial eclipses, the shadow axis misses earth entirely. Therefore, the point of greatest eclipse lies on the day/night terminator and the sun appears in the horizon.

Data pertinent to the eclipse appear with each map. In the upper left corner are the Universal Times of greatest eclipse and conjunction of the moon and sun in right ascension, the minimum distance of the moon's shadow axis from earth's center in earth radii (GAMMA) and the geocentric ratio of diameters of the moon and sun. For partial eclipses (none of which are shown here), the geocentric ratio is replaced by the magnitude at greatest eclipse. The magnitude is defined as the fraction of the sun's diameter obscured by the moon. To the upper right are exterior contact times of the moon's shadow with earth. P1 and P4 are the first and last contacts of the penumbra; they mark the start and end of the partial eclipse. Below each map are the geocentric coordinates of the sun and moon at the instant of greatest eclipse. They consist of the right ascension (RA), declination (DEC), apparent semi-diameter (SD) and horizontal parallax (HP). The saros series for the eclipse is listed, followed by a pair of numbers in parentheses. The first number identifies the sequence order of the eclipse in the saros, while the second is the total number of eclipses in the series. The Julian Date (JD) at greatest eclipse is given, followed by the extrapolated value of ΔT used in the calculations (ΔT is the difference between Terrestrial Dynamical Time and Universal Time, and Julian Days are consecutively numbered days in a series whose origin is January 1, 4713 BC).

Umbral eclipse path tables†

Tables 5.1 to 5.9 list the geodetic coordinates of the path of totality or annularity for every central eclipse in the 1990s. Appearing at the top of

† Extracted from: *Fifty Year Canon of Solar Eclipses: 1986–2035* (F. Espenak, 1987).

Table 5.1. *Annular solar eclipse of Tuesday, May 10, 1994*

Saros 128 $\Delta T = 59.5$ sec

Universal Time	Northern limit		Southern limit		Center line		Diameter ratio	Sun alt.	Sun az.	Path width	Duration annularity
	Latitude	Longitude	Latitude	Longitude	Latitude	Longitude					
Limits	14 48.0	146 54.9	12 20.3	145 22.4	13 33.8	146 8.1	0.9296	0.0	71.8	310.6	4:34.0
15:25	15 56.9	143 25.2	16 35.6	133 53.8	16 37.6	137 28.4	0.9319	9.3	74.2	303.1	4:44.2
15:30	20 17.8	132 7.2	19 26.7	127 27.4	19 54.1	129 39.1	0.9342	18.4	77.3	294.3	4:55.7
15:35	22 41.6	126 48.4	21 33.8	123 6.2	22 8.3	124 53.1	0.9356	24.4	79.9	287.7	5:3.9
15:40	24 36.9	122 51.5	23 20.7	119 38.7	23 59.1	121 12.2	0.9367	29.2	82.4	281.9	5:10.9
15:45	26 16.7	119 36.0	24 55.3	116 42.1	25 36.1	118 6.8	0.9376	33.2	84.9	276.5	5:17.2
15:50	27 46.3	116 46.0	26 21.0	114 5.9	27 3.7	115 24.1	0.9384	36.9	87.5	271.6	5:23.0
15:55	29 8.3	114 13.3	27 40.0	111 44.0	28 24.1	112 57.1	0.9391	40.2	90.3	267.0	5:28.4
16:00	30 24.5	111 52.9	28 53.5	109 32.5	29 38.9	110 41.3	0.9397	43.2	93.2	262.7	5:33.4
16:05	31 35.7	109 41.7	30 2.5	107 28.9	30 48.9	108 34.0	0.9402	46.0	96.3	258.7	5:38.2
16:10	32 42.8	107 37.2	31 7.4	105 31.2	31 54.9	106 33.0	0.9406	48.6	99.6	255.0	5:42.6
16:15	33 46.2	105 37.7	32 8.8	103 38.1	32 57.3	104 36.8	0.9411	51.1	103.1	251.5	5:46.8
16:20	34 46.3	103 42.1	33 6.9	101 48.3	33 56.4	102 44.2	0.9414	53.3	106.9	248.3	5:50.8
16:25	35 43.3	101 49.2	34 2.1	100 1.1	34 52.5	100 54.2	0.9417	55.4	111.1	245.4	5:54.4
16:30	36 37.6	99 58.3	34 54.6	98 15.6	35 45.9	99 6.0	0.9420	57.4	115.5	242.7	5:57.8
16:35	37 29.2	98 8.4	35 44.4	96 31.2	36 36.5	97 19.0	0.9423	59.2	120.4	240.3	6:0.9
16:40	38 18.3	96 19.2	36 31.6	94 47.3	37 24.7	95 32.5	0.9425	60.7	125.7	238.2	6:3.7
16:45	39 4.9	94 29.9	37 16.4	93 3.6	38 10.4	93 46.1	0.9427	62.1	131.3	236.2	6:6.1
16:50	39 49.1	92 40.2	37 58.8	91 19.5	38 53.7	91 59.2	0.9428	63.3	137.4	234.6	6:8.3
16:55	40 30.9	90 49.6	38 38.9	89 34.8	39 34.6	90 11.6	0.9430	64.3	144.0	233.1	6:10.1
17:00	41 10.3	88 57.7	39 16.5	87 48.9	40 13.1	88 22.8	0.9430	65.0	150.8	231.9	6:11.5
17:05	41 47.2	87 4.2	39 51.8	86 1.6	40 49.2	86 32.4	0.9431	65.5	158.0	231.0	6:12.6
17:10	42 21.8	85 8.6	40 24.7	84 12.6	41 22.9	84 40.2	0.9431	65.7	165.4	230.3	6:13.3
17:15	42 53.8	83 10.6	40 55.2	82 21.6	41 54.1	82 45.8	0.9431	65.6	172.8	229.7	6:13.7
17:20	43 23.3	81 9.9	41 23.1	80 28.2	42 22.8	80 48.8	0.9431	65.3	180.2	229.5	6:13.6
17:25	43 50.1	79 6.3	41 48.5	78 32.2	42 48.9	78 49.1	0.9431	64.7	187.5	229.4	6:13.1
17:30	44 14.2	76 59.2	42 11.3	76 33.3	43 12.4	76 46.2	0.9430	63.9	194.5	229.6	6:12.3
17:35	44 35.5	74 48.5	42 31.2	74 31.1	43 32.9	74 39.8	0.9429	62.8	201.2	230.0	6:11.0
17:40	44 53.7	72 33.7	42 48.3	72 25.4	43 50.6	72 29.7	0.9427	61.5	207.6	230.7	6:9.3
17:45	45 8.7	70 14.5	43 2.4	70 15.7	44 5.1	70 15.4	0.9425	60.0	213.7	231.6	6:7.2
17:50	45 20.3	67 50.4	43 13.3	68 1.8	44 16.4	67 56.6	0.9423	58.4	219.4	232.8	6:4.7
17:55	45 28.4	65 21.0	43 20.8	65 43.2	44 24.2	65 32.7	0.9421	56.5	224.8	234.3	6:1.8
18:00	45 32.6	62 45.8	43 24.7	63 19.3	44 28.2	63 3.4	0.9418	54.5	230.0	236.0	5:58.5
18:05	45 32.6	60 4.0	43 24.7	60 49.7	44 28.3	60 27.9	0.9415	52.3	235.0	238.0	5:54.8
18:10	45 28.1	57 15.0	43 20.4	58 13.5	44 23.9	57 45.5	0.9411	50.0	239.7	240.4	5:50.7
18:15	45 18.5	54 17.5	43 11.5	55 29.9	44 14.7	54 55.2	0.9407	47.5	244.3	243.1	5:46.2
18:20	45 3.3	51 10.3	42 57.5	52 37.7	44 0.1	51 55.8	0.9402	44.8	248.7	246.2	5:41.3
18:25	44 41.7	47 51.4	42 37.6	49 35.4	43 39.4	48 45.5	0.9397	41.9	253.1	249.8	5:36.0
18:30	44 12.6	44 18.2	42 11.0	46 20.7	43 11.6	45 21.9	0.9391	38.7	257.4	253.9	5:30.3
18:35	43 34.5	40 26.6	41 36.4	42 50.5	42 35.4	41 41.6	0.9384	35.3	261.7	258.6	5:24.1
18:40	42 44.9	36 10.0	40 52.0	38 59.8	41 48.6	37 38.7	0.9376	31.5	266.0	264.0	5:17.4
18:45	41 39.6	31 16.7	39 54.7	34 40.6	40 47.6	33 3.6	0.9366	27.1	270.5	270.3	5:10.0
18:50	40 9.9	25 21.0	38 38.8	29 36.9	39 25.6	27 36.5	0.9354	21.9	275.4	278.0	5:1.6
18:55	37 48.1	17 0.8	36 50.9	23 10.2	37 23.9	20 23.2	0.9337	15.1	281.0	288.2	4:51.3
Limits	33 29.1	3 13.8	31 5.2	5 1.9	32 16.8	4 8.8	0.9299	0.0	291.1	309.0	4:32.2

Eclipse predictions and map courtesy of F. Espenak – NASA/GSFC.

Table 5.2. *Total solar eclipse of Thursday, November 3, 1994*

Saros 133 $\Delta T = 59.7$ sec

Universal Time	Northern limit Latitude	Northern limit Longitude	Southern limit Latitude	Southern limit Longitude	Center line Latitude	Center line Longitude	Diameter ratio	Sun alt.	Sun az.	Path width	Duration totality
Limits	−7 −19.8	96 38.0	−8 −27.1	97 11.0	−7 −53.4	96 54.6	1.0360	0.0	105.2	135.3	1:51.9
12:05	−11 −2.9	84 36.7	−11 −25.1	87 40.2	−11 −15.6	86 4.0	1.0397	11.7	103.3	150.7	2:13.9
12:10	−13 −51.0	77 18.6	−14 −34.9	79 23.1	−14 −13.5	78 19.6	1.0426	20.9	101.0	162.0	2:33.2
12:15	−15 −56.8	72 28.0	−16 −49.0	74 16.1	−16 −23.3	73 21.3	1.0445	27.2	98.8	169.0	2:47.3
12:20	−17 −44.0	68 38.5	−18 −41.6	70 18.2	−18 −13.1	69 27.7	1.0459	32.3	96.7	174.0	2:59.2
12:25	−19 −19.6	65 24.0	−20 −21.4	66 58.1	−19 −50.8	66 10.4	1.0471	36.8	94.3	177.9	3:9.7
12:30	−20 −47.2	62 32.1	−21 −52.5	64 2.2	−21 −20.1	63 16.6	1.0481	40.7	91.9	180.9	3:19.2
12:35	−22 −8.5	59 56.2	−23 −16.9	61 23.0	−22 −42.9	60 39.1	1.0490	44.3	89.2	183.3	3:27.9
12:40	−23 −24.9	57 31.9	−24 −36.1	58 55.8	−24 −0.7	58 13.4	1.0497	47.6	86.2	185.2	3:35.9
12:45	−24 −37.0	55 16.3	−25 −50.8	56 37.5	−25 −14.1	55 56.5	1.0504	50.7	83.0	186.6	3:43.2
12:50	−25 −45.5	53 7.2	−27 −1.8	54 25.9	−26 −23.7	53 46.1	1.0510	53.6	79.4	187.7	3:49.8
12:55	−26 −50.7	51 3.0	−28 −9.3	52 19.1	−27 −30.1	51 40.7	1.0515	56.3	75.5	188.6	3:55.9
13:00	−27 −53.0	49 2.4	−29 −13.7	50 16.0	−28 −33.4	49 38.8	1.0519	58.7	71.0	189.2	4:1.4
13:05	−28 −52.5	47 4.5	−30 −15.4	48 15.3	−29 −34.0	47 39.5	1.0523	61.0	66.0	189.6	4:6.3
13:10	−29 −49.4	45 8.2	−31 −14.3	46 16.1	−30 −31.9	45 41.8	1.0526	63.1	60.4	189.9	4:10.5
13:15	−30 −43.9	43 12.8	−32 −10.8	44 17.7	−31 −27.4	43 44.9	1.0529	64.9	54.2	190.0	4:14.2
13:20	−31 −36.1	41 17.8	−33 −4.7	42 19.4	−32 −20.4	41 48.3	1.0531	66.4	47.2	190.0	4:17.3
13:25	−32 −25.9	39 22.3	−33 −56.3	40 20.5	−33 −11.1	39 51.1	1.0533	67.6	39.5	189.9	4:19.8
13:30	−33 −13.4	37 26.0	−34 −45.4	38 20.4	−33 −59.4	37 53.0	1.0534	68.5	31.1	189.6	4:21.6
13:35	−33 −58.6	35 28.3	−35 −32.2	36 18.7	−34 −45.4	35 53.3	1.0535	69.1	22.2	189.3	4:22.8
13:40	−34 −41.5	33 28.6	−36 −16.5	34 14.6	−35 −29.0	33 51.4	1.0535	69.2	13.0	188.9	4:23.3
13:45	−35 −22.1	31 26.4	−36 −58.4	32 7.8	−36 −10.2	31 47.0	1.0535	68.9	3.8	188.4	4:23.2
13:50	−36 −0.2	29 21.3	−37 −37.7	29 57.6	−36 −48.9	29 39.4	1.0534	68.2	354.9	187.9	4:22.4
13:55	−36 −35.9	27 12.6	−38 −14.3	27 43.6	−37 −25.0	27 28.1	1.0533	67.2	346.5	187.2	4:21.0
14:00	−37 −8.9	24 59.9	−38 −48.1	25 25.1	−37 −58.4	25 12.6	1.0531	65.8	338.6	186.5	4:18.9
14:05	−37 −39.1	22 42.6	−39 −18.9	23 1.6	−38 −28.9	22 52.2	1.0529	64.2	331.4	185.6	4:16.1
14:10	−38 −6.4	20 19.9	−39 −46.5	20 32.4	−38 −56.4	20 26.3	1.0526	62.3	324.7	184.7	4:12.6
14:15	−38 −30.6	17 51.3	−40 −10.7	17 56.7	−39 −20.6	17 54.2	1.0523	60.1	318.6	183.7	4:8.4
14:20	−38 −51.3	15 15.7	−40 −31.2	15 13.7	−39 −41.2	15 15.1	1.0519	57.8	312.9	182.5	4:3.6
14:25	−39 −8.3	12 32.4	−40 −47.7	12 22.4	−39 −57.9	12 27.8	1.0515	55.2	307.6	181.2	3:58.1
14:30	−39 −21.2	9 39.9	−40 −59.6	9 21.6	−40 −10.4	9 31.3	1.0509	52.5	302.7	179.8	3:51.9
14:35	−39 −29.4	6 37.0	−41 −6.5	6 9.6	−40 −17.9	6 24.0	1.0503	49.5	297.9	178.2	3:45.0
14:40	−39 −32.4	3 21.7	−41 −7.5	2 44.6	−40 −20.0	3 3.9	1.0496	46.4	293.3	176.4	3:37.3
14:45	−39 −29.3	0 −8.8	−41 −1.9	0 −56.2	−40 −15.6	0 −31.6	1.0489	43.0	288.8	174.3	3:28.9
14:50	−39 −18.8	−3 −57.9	−40 −48.1	−4 −56.8	−40 −3.5	−4 −26.4	1.0479	39.2	284.3	171.9	3:19.6
14:55	−38 −59.3	−8 −11.4	−40 −24.2	−9 −23.2	−39 −41.9	−8 −46.2	1.0469	35.1	279.8	169.1	3:9.3
15:00	−38 −27.8	−12 −58.6	−39 −46.9	−14 −25.9	−39 −7.6	−13 −40.9	1.0456	30.5	275.1	165.6	2:57.9
15:05	−37 −39.1	−18 −37.2	−38 −49.9	−20 −25.3	−38 −14.9	−19 −29.6	1.0441	25.0	270.1	161.2	2:44.8
15:10	−36 −20.8	−25 −50.4	−37 −17.2	−28 −15.1	−36 −49.8	−27 −0.1	1.0420	18.0	264.3	155.0	2:28.7
Limits	−31 −24.9	−46 −25.3	−32 −31.7	−47 −5.0	−31 −58.3	−46 −45.0	1.0362	0.0	252.1	136.4	1:52.7

Eclipse predictions and map courtesy of F. Espenak – NASA/GSFC.

Table 5.3. *Annular solar eclipse of Saturday, April 29, 1995*

Saros 138 $\Delta T = 60.0$ sec

Universal Time	Northern limit		Southern limit		Center line		Diameter ratio	Sun alt.	Sun az.	Path width	Duration annularity
	Latitude	Longitude	Latitude	Longitude	Latitude	Longitude					
Limits	−30 −36.4	137 16.9	−32 −46.3	136 45.5	−31 −41.2	137 1.2	0.9361	0.0	72.9	244.0	4:32.3
15:45	−26 −32.7	123 49.1	−29 −6.8	124 48.2	−27 −47.5	124 13.2	0.9392	12.4	66.2	229.5	4:47.6
15:50	−23 −56.9	116 56.7	−26 −8.1	116 56.9	−25 −1.6	116 55.2	0.9412	20.4	62.1	220.8	4:59.2
15:55	−21 −59.8	112 22.8	−24 −1.3	112 4.5	−22 −59.9	112 12.7	0.9426	26.2	59.3	215.1	5:8.3
16:00	−20 −21.0	108 50.6	−22 −16.3	108 22.9	−21 −18.1	108 36.1	0.9437	30.9	56.9	210.8	5:16.3
16:05	−18 −53.6	105 54.9	−20 −44.4	105 21.6	−19 −48.6	105 37.8	0.9445	35.0	54.8	207.3	5:23.5
16:10	−17 −34.5	103 23.8	−19 −21.8	102 46.8	−18 −27.8	103 4.9	0.9453	38.7	52.8	204.5	5:30.3
16:15	−16 −21.6	101 10.4	−18 −6.2	100 30.8	−17 −13.6	100 50.3	0.9459	42.1	50.8	202.0	5:36.7
16:20	−15 −13.7	99 10.4	−16 −56.2	98 29.0	−16 −4.7	98 49.5	0.9465	45.2	48.8	200.0	5:42.8
16:25	−14 −10.1	97 20.9	−15 −50.8	96 38.2	−15 −0.2	96 59.4	0.9470	48.1	46.7	198.3	5:48.6
16:30	−13 −10.2	95 39.7	−14 −49.4	94 56.2	−13 −59.5	95 17.8	0.9475	50.9	44.4	196.8	5:54.1
16:35	−12 −13.4	94 5.3	−13 −51.5	93 21.1	−13 −2.2	93 43.1	0.9479	53.4	42.0	195.7	5:59.3
16:40	−11 −19.5	92 36.4	−12 −56.6	91 51.9	−12 −7.8	92 14.0	0.9482	55.9	39.4	194.7	6:4.3
16:45	−10 −28.1	91 12.1	−12 −4.6	90 27.3	−11 −16.1	90 49.6	0.9485	58.2	36.5	194.0	6:9.0
16:50	−9 −39.1	89 51.5	−11 −15.1	89 6.6	−10 −26.8	89 29.0	0.9488	60.3	33.3	193.5	6:13.4
16:55	−8 −52.4	88 34.0	−10 −27.9	87 49.1	−9 −39.9	88 11.5	0.9490	62.3	29.7	193.2	6:17.5
17:00	−8 −7.6	87 19.0	−9 −43.0	86 34.2	−8 −55.1	86 56.5	0.9492	64.1	25.7	193.0	6:21.3
17:05	−7 −24.9	86 6.0	−9 −0.2	85 21.5	−8 −12.3	85 43.7	0.9494	65.7	21.2	193.0	6:24.8
17:10	−6 −44.0	84 54.6	−8 −19.4	84 10.3	−7 −31.4	84 32.4	0.9495	67.1	16.1	193.2	6:27.9
17:15	−6 −4.9	83 44.4	−7 −40.6	83 0.4	−6 −52.5	83 22.4	0.9496	68.3	10.4	193.5	6:30.6
17:20	−5 −27.5	82 35.0	−7 −3.5	81 51.4	−6 −15.3	82 13.2	0.9497	69.2	4.3	193.9	6:32.9
17:25	−4 −51.9	81 26.1	−6 −28.3	80 43.0	−5 −39.9	81 4.5	0.9497	69.8	357.7	194.5	6:34.8
17:30	−4 −17.9	80 17.3	−5 −54.9	79 34.7	−5 −6.2	79 56.0	0.9497	70.2	350.8	195.2	6:36.3
17:35	−3 −45.6	79 8.4	−5 −23.3	78 26.4	−4 −34.2	78 47.4	0.9497	70.2	343.8	195.9	6:37.3
17:40	−3 −15.0	77 59.1	−4 −53.4	77 17.7	−4 −3.9	77 38.3	0.9497	69.8	337.0	196.7	6:37.8
17:45	−2 −46.0	76 49.0	−4 −25.2	76 8.2	−3 −35.3	76 28.6	0.9496	69.2	330.5	197.6	6:37.8
17:50	−2 −18.8	75 37.8	−3 −58.8	74 57.7	−3 −8.5	75 17.8	0.9495	68.3	324.4	198.6	6:37.3
17:55	−1 −53.2	74 25.3	−3 −34.2	73 45.9	−2 −43.4	74 5.6	0.9493	67.1	319.0	199.6	6:36.3
18:00	−1 −29.4	73 11.0	−3 −11.4	72 32.3	−2 −20.1	72 51.7	0.9492	65.6	314.2	200.6	6:34.7
18:05	−1 −7.4	71 54.6	−2 −50.6	71 16.7	−1 −58.7	71 35.7	0.9490	64.0	309.9	201.6	6:32.7
18:10	0 −47.4	70 35.7	−2 −31.7	69 58.5	−1 −39.2	70 17.2	0.9487	62.2	306.2	202.7	6:30.1
18:15	0 −29.3	69 13.8	−2 −14.8	68 37.4	−1 −21.8	68 55.7	0.9485	60.2	303.0	203.7	6:26.9
18:20	0 −13.3	67 48.5	−2 −0.1	67 12.8	−1 −6.4	67 30.7	0.9482	58.1	300.3	204.9	6:23.2
18:25	0 0.4	66 19.0	−1 −47.8	65 44.1	0 −53.4	66 1.6	0.9478	55.8	297.9	206.0	6:19.0
18:30	0 11.7	64 44.7	−1 −38.0	64 10.4	0 −42.9	64 27.7	0.9474	53.4	295.8	207.2	6:14.3
18:35	0 20.3	63 4.7	−1 −30.9	62 30.9	0 −35.0	62 48.0	0.9470	50.8	294.1	208.5	6:9.1
18:40	0 26.0	61 17.9	−1 −27.0	60 44.5	0 −30.2	61 1.4	0.9465	48.0	292.6	209.8	6:3.4
18:45	0 28.3	59 22.8	−1 −26.5	58 49.6	0 −28.8	59 6.4	0.9460	45.1	291.3	211.3	5:57.2
18:50	0 26.8	57 17.5	−1 −30.1	56 44.2	0 −31.3	57 1.1	0.9454	42.0	290.2	213.0	5:50.5
18:55	0 20.7	54 59.5	−1 −38.5	54 25.5	0 −38.6	54 42.8	0.9447	38.6	289.2	215.0	5:43.3
19:00	0 9.0	52 24.8	−1 −53.0	51 49.2	0 −51.6	52 7.5	0.9439	34.9	288.4	217.3	5:35.5
19:05	0 −9.9	49 27.2	−2 −15.4	48 48.8	−1 −12.2	49 8.6	0.9430	30.7	287.7	220.1	5:27.0
19:10	0 −39.1	45 56.1	−2 −49.1	45 11.8	−1 −43.6	45 34.9	0.9419	26.0	287.1	223.5	5:17.6
19:15	−1 −24.5	41 28.4	−3 −41.7	40 30.8	−2 −32.4	41 1.2	0.9405	20.2	286.5	228.1	5:6.8
19:20	−2 −44.8	34 52.8	−5 −20.4	33 5.5	−4 −0.8	34 4.3	0.9384	11.9	285.7	235.4	4:52.8
Limits	−5 −36.8	22 54.6	−7 −50.2	23 14.2	−6 −43.3	23 4.6	0.9353	0.0	284.6	246.9	4:35.8

Eclipse predictions and map courtesy of F. Espenak – NASA/GSFC.

Table 5.4. *Total solar eclipse of Tuesday, October 24, 1995*

Saros 143 ΔT = 60.4 sec

Universal Time	Northern limit Latitude	Northern limit Longitude	Southern limit Latitude	Southern limit Longitude	Center line Latitude	Center line Longitude	Diameter ratio	Sun alt.	Sun az.	Path width	Duration totality
Limits	34 54.7	−51 −6.4	34 46.2	−51 −4.9	34 50.5	−51 −5.7	1.0044	0.0	104.1	15.8	0:15.8
2:55	31 51.1	−63 −47.3	31 31.2	−64 −0.6	31 41.2	−63 −54.1	1.0079	11.6	111.5	28.1	0:31.0
3:00	29 5.0	−72 −34.9	28 42.0	−72 −38.5	28 53.5	−72 −36.8	1.0105	20.6	116.9	37.3	0:44.5
3:05	27 3.1	−78 −2.6	26 37.8	−78 −1.7	26 50.5	−78 −2.3	1.0123	26.8	120.4	43.3	0:54.5
3:10	25 20.0	−82 −11.5	24 52.9	−82 −7.6	25 6.4	−82 −9.6	1.0136	31.8	123.4	48.0	1:3.1
3:15	23 48.2	−85 −35.7	23 19.7	−85 −29.3	23 33.9	−85 −32.6	1.0148	36.1	126.1	51.9	1:10.8
3:20	22 24.5	−88 −30.4	21 54.7	−88 −22.1	22 9.6	−88 −26.3	1.0157	40.0	128.6	55.2	1:17.8
3:25	21 6.9	−91 −4.1	20 36.1	−90 −54.2	20 51.5	−90 −59.2	1.0166	43.6	131.2	58.2	1:24.3
3:30	19 54.3	−93 −22.2	19 22.5	−93 −10.9	19 38.3	−93 −16.6	1.0173	46.8	133.7	60.8	1:30.3
3:35	18 45.7	−95 −28.2	18 13.1	−95 −15.7	18 29.4	−95 −22.0	1.0180	49.9	136.4	63.1	1:35.8
3:40	17 40.7	−97 −24.6	17 7.3	−97 −11.1	17 24.0	−97 −17.9	1.0185	52.7	139.3	65.3	1:40.9
3:45	16 38.8	−99 −13.4	16 4.7	−98 −58.9	16 21.7	−99 −6.2	1.0191	55.4	142.4	67.2	1:45.6
3:50	15 39.6	−100 −55.9	15 4.9	−100 −40.6	15 22.3	−100 −48.3	1.0195	57.8	145.8	68.9	1:49.9
3:55	14 43.0	−102 −33.3	14 7.7	−102 −17.4	14 25.3	−102 −25.4	1.0199	60.1	149.6	70.5	1:53.8
4:00	13 48.7	−104 −6.7	13 12.9	−103 −50.2	13 30.8	−103 −58.5	1.0202	62.2	153.9	71.9	1:57.3
4:05	12 56.5	−105 −36.7	12 20.3	−105 −19.8	12 38.4	−105 −28.3	1.0205	64.1	158.6	73.1	2:0.4
4:10	12 6.4	−107 −4.1	11 29.7	−106 −46.9	11 48.0	−106 −55.5	1.0208	65.7	164.0	74.2	2:3.1
4:15	11 18.2	−108 −29.5	10 41.1	−108 −12.0	10 59.6	−108 −20.8	1.0210	67.1	169.9	75.2	2:5.3
4:20	10 31.8	−109 −53.5	9 54.4	−109 −35.7	10 13.1	−109 −44.6	1.0211	68.2	176.5	76.1	2:7.0
4:25	9 47.2	−111 −16.4	9 9.5	−110 −58.5	9 28.4	−111 −7.5	1.0212	68.9	183.5	76.8	2:8.4
4:30	9 4.4	−112 −38.9	8 26.4	−112 −20.9	8 45.4	−112 −29.9	1.0213	69.3	190.9	77.3	2:9.2
4:35	8 23.2	−114 −1.2	7 45.1	−113 −43.2	8 4.2	−113 −52.2	1.0214	69.3	198.4	77.7	2:9.6
4:40	7 43.8	−115 −24.0	7 5.5	−115 −6.0	7 24.6	−115 −15.0	1.0213	68.9	205.8	78.0	2:9.5
4:45	7 6.0	−116 −47.5	6 27.6	−116 −29.6	6 46.8	−116 −38.6	1.0213	68.2	212.7	78.1	2:9.0
4:50	6 29.9	−118 −12.3	5 51.5	−117 −54.5	6 10.7	−118 −3.4	1.0212	67.1	219.1	78.0	2:7.9
4:55	5 55.6	−119 −38.7	5 17.2	−119 −21.2	5 36.4	−119 −30.0	1.0211	65.8	224.8	77.7	2:6.4
5:00	5 23.0	−121 −7.4	4 44.7	−120 −50.1	5 3.9	−120 −58.7	1.0209	64.1	229.9	77.2	2:4.4
5:05	4 52.3	−122 −38.8	4 14.1	−122 −21.8	4 33.2	−122 −30.3	1.0207	62.3	234.3	76.5	2:1.9
5:10	4 23.5	−124 −13.6	3 45.6	−123 −57.0	4 4.5	−124 −5.3	1.0204	60.2	238.1	75.6	1:58.9
5:15	3 56.7	−125 −52.5	3 19.1	−125 −36.2	3 37.9	−125 −44.3	1.0201	57.9	241.3	74.4	1:55.4
5:20	3 32.0	−127 −36.2	2 54.9	−127 −20.4	3 13.5	−127 −28.3	1.0197	55.4	244.1	73.0	1:51.4
5:25	3 9.8	−129 −25.9	2 33.2	−129 −10.5	2 51.5	−129 −18.2	1.0192	52.8	246.5	71.3	1:47.0
5:30	2 50.1	−131 −22.8	2 14.3	−131 −7.8	2 32.2	−131 −15.2	1.0187	50.0	248.6	69.4	1:42.1
5:35	2 33.4	−133 −28.4	1 58.4	−133 −13.8	2 15.9	−133 −21.1	1.0181	46.9	250.3	67.1	1:36.6
5:40	2 20.0	−135 −45.0	1 46.0	−135 −30.7	2 3.0	−135 −37.8	1.0175	43.7	251.8	64.5	1:30.7
5:45	2 10.6	−138 −15.4	1 37.8	−138 −1.5	1 54.3	−138 −8.4	1.0167	40.1	253.1	61.4	1:24.2
5:50	2 6.2	−141 −4.1	1 34.8	−140 −50.5	1 50.5	−140 −57.2	1.0158	36.2	254.2	57.9	1:17.1
5:55	2 8.2	−144 −18.2	1 38.5	−144 −4.6	1 53.4	−144 −11.3	1.0147	31.9	255.1	53.8	1:9.3
6:00	2 19.3	−148 −10.2	1 51.7	−147 −56.3	2 5.5	−148 −3.1	1.0134	26.9	255.9	48.9	1:0.5
6:05	2 45.2	−153 −7.8	2 20.3	−152 −52.8	2 32.8	−153 −0.2	1.0118	20.8	256.6	42.6	0:50.1
6:10	3 46.2	−160 −48.3	3 24.7	−160 −28.1	3 35.5	−160 −38.1	1.0092	11.9	257.3	33.2	0:36.4
Limits	5 44.9	−171 −47.1	5 33.8	−171 −47.9	5 39.4	−171 −47.5	1.0056	0.0	258.3	20.3	0:20.2

Eclipse predictions and map courtesy of F. Espenak – NASA/GSFC.

Table 5.5. *Total solar eclipse of Sunday, March 9, 1997*

Saros 120 $\Delta T = 61.4$ sec

Universal Time	Northern limit Latitude	Northern limit Longitude	Southern limit Latitude	Southern limit Longitude	Center line Latitude	Center line Longitude	Diameter ratio	Sun alt.	Sun az.	Path width	Duration totality
Limits	50 50.6	−86 −50.1	48 8.4	−87 −26.9	49 27.2	−87 −8.8	1.0349	0.0	97.0	290.9	1:58.1
0:50	50 50.1	−101 −5.8	48 29.5	−108 −28.3	49 38.2	−105 −11.7	1.0388	12.3	112.4	346.3	2:24.4
0:55	51 40.0	−107 −37.4	49 22.5	−113 −19.8	50 30.3	−110 −41.4	1.0399	15.8	118.3	361.0	2:32.8
1:00	52 41.5	−112 −24.6	50 23.0	−117 −19.7	51 31.5	−115 −1.3	1.0407	18.3	123.4	368.6	2:38.9
1:05	53 50.5	−116 −24.4	51 29.6	−120 −50.6	52 39.3	−118 −45.0	1.0412	20.1	128.3	371.2	2:43.4
1:10	55 5.9	−119 −57.7	52 41.5	−124 −4.1	53 52.9	−122 −7.5	1.0416	21.4	133.0	370.2	2:46.6
1:15	56 27.5	−123 −15.7	53 58.8	−127 −7.3	55 12.2	−125 −17.6	1.0419	22.3	137.7	366.5	2:48.8
1:20	57 55.5	−126 −25.6	55 21.6	−130 −5.3	56 37.4	−128 −21.4	1.0420	22.8	142.5	361.1	2:49.9
1:25	59 30.5	−129 −33.2	56 50.3	−133 −2.6	58 9.2	−131 −23.8	1.0420	22.8	147.2	354.4	2:50.1
1:30	61 13.7	−132 −43.9	58 25.9	−136 −3.5	59 48.2	−134 −29.7	1.0419	22.5	152.1	347.1	2:49.2
1:35	63 6.6	−136 −3.8	60 9.3	−139 −13.0	61 36.1	−137 −44.5	1.0417	21.9	157.2	339.5	2:47.4
1:40	65 12.0	−139 −40.9	62 2.3	−142 −37.3	63 34.7	−141 −15.3	1.0413	20.8	162.5	331.8	2:44.6
1:45	67 34.0	−143 −47.9	64 7.5	−146 −25.6	65 47.4	−145 −13.1	1.0409	19.2	168.2	324.2	2:40.5
1:50	70 20.8	−148 −49.2	66 29.2	−150 −53.5	68 20.0	−149 −57.0	1.0402	17.0	174.6	316.8	2:35.1
1:55	73 50.8	−155 −46.9	69 15.5	−156 −32.2	71 24.2	−156 −9.6	1.0392	14.0	182.3	309.5	2:27.9
2:00	79 19.2	−170 −3.8	72 45.5	−164 −44.9	75 32.9	−166 −14.0	1.0378	9.5	193.7	301.9	2:17.4
Limits	83 57.7	166 15.9	81 55.3	152 47.3	82 58.1	158 17.4	1.0347	0.0	229.8	291.9	1:57.4

Eclipse predictions and map courtesy of F. Espenak – NASA/GSFC.

Table 5.6. *Total solar eclipse of Thursday, February 26, 1998*

Saros 130 ΔT = 62.2 sec

Universal Time	Northern limit Latitude	Northern limit Longitude	Southern limit Latitude	Southern limit Longitude	Center line Latitude	Center line Longitude	Diameter ratio	Sun alt.	Sun az.	Path width	Duration totality
Limits	−1 −57.1	143 56.5	−2 −45.6	144 3.9	−2 −21.4	144 0.6	1.0259	0.0	98.6	89.4	1:27.3
15:50	−3 −19.7	132 42.2	−4 −20.0	131 58.4	−3 −49.9	132 19.3	1.0298	12.4	98.0	102.9	1:50.5
15:55	−3 −52.1	124 38.6	−4 −54.7	124 7.8	−4 −23.3	124 22.7	1.0326	21.5	97.6	112.9	2:10.3
16:00	−3 −57.4	119 26.9	−5 −2.2	118 58.6	−4 −29.8	119 12.3	1.0345	27.9	97.4	119.6	2:25.1
16:05	−3 −51.2	115 24.2	−4 −57.9	114 56.4	−4 −24.5	115 9.9	1.0359	33.1	97.4	124.9	2:37.8
16:10	−3 −38.0	112 1.4	−4 −46.2	111 33.3	−4 −12.1	111 47.1	1.0372	37.7	97.7	129.4	2:49.2
16:15	−3 −20.0	109 5.3	−4 −29.4	108 36.5	−3 −54.6	108 50.7	1.0382	41.8	98.1	133.2	2:59.5
16:20	−2 −58.1	106 28.5	−4 −8.5	105 58.9	−3 −33.3	106 13.4	1.0391	45.5	98.7	136.5	3:8.9
16:25	−2 −33.3	104 6.3	−3 −44.5	103 35.8	−3 −8.8	103 50.8	1.0399	49.1	99.5	139.4	3:17.6
16:30	−2 −5.8	101 55.6	−3 −17.7	101 24.2	−2 −41.7	101 39.7	1.0406	52.4	100.6	141.9	3:25.5
16:35	−1 −36.2	99 54.3	−2 −48.5	99 21.9	−2 −12.3	99 37.9	1.0412	55.5	102.0	144.0	3:32.8
16:40	−1 −4.5	98 0.5	−2 −17.3	97 27.3	−1 −40.9	97 43.7	1.0417	58.4	103.8	145.9	3:39.3
16:45	0 −31.1	96 13.0	−1 −44.1	95 38.9	−1 −7.6	95 55.8	1.0422	61.2	106.0	147.4	3:45.2
16:50	0 3.9	94 30.7	−1 −9.3	93 55.9	0 −32.7	94 13.2	1.0426	63.9	108.7	148.7	3:50.5
16:55	0 40.5	92 52.7	0 −32.8	92 17.2	0 3.9	92 34.9	1.0430	66.4	112.1	149.7	3:55.1
17:00	1 18.5	91 18.4	0 5.2	90 42.2	0 41.8	91 0.2	1.0433	68.7	116.3	150.5	3:59.0
17:05	1 57.9	89 47.0	0 44.6	89 10.2	1 21.3	89 28.5	1.0435	70.8	121.6	151.1	4:2.2
17:10	2 38.6	88 17.9	1 25.4	87 40.7	2 2.0	87 59.3	1.0437	72.7	128.1	151.4	4:4.8
17:15	3 20.6	86 50.8	2 7.6	86 13.1	2 44.1	86 31.9	1.0439	74.2	136.1	151.6	4:6.7
17:20	4 4.0	85 24.9	2 51.0	84 47.0	3 27.5	85 5.9	1.0440	75.4	145.5	151.7	4:7.9
17:25	4 48.5	83 60.0	3 35.8	83 21.8	4 12.2	83 40.9	1.0441	76.0	156.2	151.5	4:8.5
17:30	5 34.4	82 35.5	4 21.9	81 57.2	4 58.1	82 16.4	1.0441	76.1	167.6	151.2	4:8.5
17:35	6 21.6	81 11.0	5 9.3	80 32.7	5 45.4	80 51.9	1.0441	75.6	178.7	150.8	4:7.8
17:40	7 10.0	79 46.1	5 58.0	79 7.8	6 34.0	79 27.0	1.0440	74.7	188.7	150.2	4:6.4
17:45	7 59.8	78 20.2	6 48.1	77 42.1	7 23.9	78 1.2	1.0439	73.2	197.4	149.5	4:4.4
17:50	8 51.0	76 53.0	7 39.5	76 15.1	8 15.2	76 34.1	1.0438	71.5	204.7	148.7	4:1.8
17:55	9 43.6	75 23.7	8 32.5	74 46.3	9 8.0	75 5.1	1.0436	69.4	210.7	147.8	3:58.6
18:00	10 37.8	73 52.0	9 26.9	73 15.1	10 2.3	73 33.6	1.0433	67.2	215.7	146.7	3:54.8
18:05	11 33.6	72 17.0	10 23.0	71 40.8	10 58.2	71 59.0	1.0430	64.7	219.9	145.5	3:50.4
18:10	12 31.1	70 37.9	11 20.8	70 2.6	11 55.9	70 20.4	1.0427	62.1	223.4	144.1	3:45.3
18:15	13 30.4	68 54.0	12 20.6	68 19.7	12 55.4	68 37.0	1.0423	59.4	226.5	142.6	3:39.7
18:20	14 31.8	67 4.0	13 22.3	66 30.9	13 57.0	66 47.6	1.0418	56.5	229.2	140.9	3:33.6
18:25	15 35.5	65 6.7	14 26.4	64 35.0	15 0.9	64 51.0	1.0413	53.4	231.6	139.0	3:26.8
18:30	16 41.8	63 0.1	15 33.1	62 30.2	16 7.3	62 45.3	1.0406	50.2	233.9	136.9	3:19.4
18:35	17 51.0	60 42.1	16 42.7	60 14.2	17 16.8	60 28.4	1.0399	46.8	236.0	134.6	3:11.3
18:40	19 3.7	58 9.4	17 56.0	57 44.0	18 29.7	57 57.0	1.0391	43.1	238.1	131.9	3:2.6
18:45	20 20.6	55 17.5	19 13.5	54 55.1	19 46.9	55 6.6	1.0382	39.1	240.2	128.8	2:53.1
18:50	21 43.0	51 59.1	20 36.5	51 40.6	21 9.6	51 50.2	1.0371	34.7	242.4	125.3	2:42.6
18:55	23 12.8	48 1.9	22 7.1	47 48.7	22 39.8	47 55.7	1.0358	29.8	244.8	121.1	2:30.9
19:00	24 53.9	43 1.0	23 49.2	42 55.8	24 21.4	42 58.9	1.0341	23.9	247.7	115.8	2:17.5
19:05	26 57.3	35 46.0	25 53.3	35 57.0	26 25.1	35 52.4	1.0318	16.1	251.5	108.4	2:0.6
Limits	30 15.9	19 5.9	29 26.1	19 0.3	29 51.0	19 3.1	1.0268	0.0	260.1	92.3	1:30.0

Eclipse predictions and map courtesy of F. Espenak – NASA/GSFC.

Table 5.7. *Annular solar eclipse of Saturday, August 22, 1998*

Saros 135 $\Delta T = 62.6$ sec

Universal Time	Northern limit Latitude	Northern limit Longitude	Southern limit Latitude	Southern limit Longitude	Center line Latitude	Center line Longitude	Diameter ratio	Sun alt.	Sun az.	Path width	Duration annularity
Limits	0 −0.9	−86 −54.1	−1 −22.5	−87 −4.6	0 −41.7	−86 −59.5	0.9590	0.0	78.1	151.1	2:44.3
0:20	2 37.2	−101 −31.7	1 16.6	−100 −52.8	1 57.2	−101 −12.9	0.9631	15.5	78.2	136.1	2:49.6
0:25	3 22.7	−107 −29.4	2 8.9	−107 −2.7	2 46.0	−107 −16.4	0.9649	22.8	78.2	129.6	2:52.6
0:30	3 45.7	−111 −47.1	2 35.9	−111 −24.3	3 10.9	−111 −35.9	0.9663	28.4	78.2	125.0	2:55.1
0:35	3 56.9	−115 −15.0	2 50.0	−114 −53.9	3 23.6	−115 −4.6	0.9673	33.0	78.0	121.4	2:57.4
0:40	4 0.3	−118 −11.8	2 55.9	−117 −51.5	3 28.2	−118 −1.7	0.9682	37.2	77.7	118.5	2:59.4
0:45	3 58.1	−120 −46.9	2 55.7	−120 −27.0	3 27.0	−120 −37.0	0.9690	40.9	77.2	115.9	3:1.3
0:50	3 51.5	−123 −6.0	2 50.9	−122 −46.2	3 21.3	−122 −56.1	0.9696	44.3	76.6	113.7	3:3.0
0:55	3 41.2	−125 −12.5	2 42.2	−124 −52.6	3 11.8	−125 −2.6	0.9702	47.6	75.8	111.7	3:4.6
1:00	3 27.9	−127 −9.1	2 30.2	−126 −49.0	2 59.1	−126 −59.0	0.9707	50.6	74.8	110.0	3:6.1
1:05	3 11.8	−128 −57.4	2 15.5	−128 −37.0	2 43.7	−128 −47.2	0.9712	53.4	73.6	108.4	3:7.4
1:10	2 53.3	−130 −38.9	1 58.2	−130 −18.3	2 25.9	−130 −28.6	0.9716	56.2	72.1	107.0	3:8.6
1:15	2 32.7	−132 −14.7	1 38.6	−131 −53.8	2 5.8	−132 −4.2	0.9719	58.8	70.3	105.8	3:9.7
1:20	2 10.1	−133 −45.6	1 17.0	−133 −24.4	1 43.6	−133 −35.0	0.9722	61.2	68.1	104.6	3:10.7
1:25	1 45.7	−135 −12.5	0 53.5	−134 −50.9	1 19.6	−135 −1.7	0.9725	63.5	65.4	103.6	3:11.5
1:30	1 19.5	−136 −35.8	0 28.1	−136 −13.9	0 53.9	−136 −24.8	0.9727	65.7	62.2	102.6	3:12.3
1:35	0 51.7	−137 −56.1	0 1.0	−137 −33.9	0 26.4	−137 −45.0	0.9729	67.7	58.3	101.8	3:12.9
1:40	0 22.3	−139 −14.0	0 −27.7	−138 −51.4	0 −2.6	−139 −2.7	0.9731	69.6	53.7	101.0	3:13.3
1:45	0 −8.6	−140 −29.7	0 −58.0	−140 −6.8	0 −33.2	−140 −18.3	0.9732	71.2	48.1	100.4	3:13.7
1:50	0 −40.9	−141 −43.7	−1 −29.8	−141 −20.6	−1 −5.3	−141 −32.2	0.9733	72.6	41.5	99.8	3:13.9
1:55	−1 −14.7	−142 −56.4	−2 −3.2	−142 −32.9	−1 −38.9	−142 −44.7	0.9733	73.6	34.0	99.4	3:14.1
2:00	−1 −49.9	−144 −8.1	−2 −38.0	−143 −44.3	−2 −13.9	−143 −56.2	0.9734	74.3	25.5	99.0	3:14.1
2:05	−2 −26.5	−145 −19.1	−3 −14.3	−144 −55.0	−2 −50.3	−145 −7.1	0.9734	74.6	16.4	98.8	3:14.0
2:10	−3 −4.4	−146 −29.7	−3 −52.0	−146 −5.4	−3 −28.2	−146 −17.6	0.9733	74.5	7.2	98.6	3:13.8
2:15	−3 −43.8	−147 −40.3	−4 −31.3	−147 −15.7	−4 −7.5	−147 −28.0	0.9732	74.0	358.4	98.6	3:13.6
2:20	−4 −24.6	−148 −51.2	−5 −12.0	−148 −26.4	−4 −48.3	−148 −38.8	0.9731	73.1	350.2	98.7	3:13.2
2:25	−5 −6.8	−150 −2.8	−5 −54.3	−149 −37.6	−5 −30.5	−149 −50.2	0.9730	71.8	343.1	98.9	3:12.8
2:30	−5 −50.6	−151 −15.3	−6 −38.2	−150 −49.9	−6 −14.3	−151 −2.6	0.9728	70.3	336.8	99.3	3:12.3
2:35	−6 −35.9	−152 −29.2	−7 −23.7	−152 −3.6	−6 −59.7	−152 −16.4	0.9727	68.5	331.5	99.8	3:11.7
2:40	−7 −22.7	−153 −44.9	−8 −10.9	−153 −19.1	−7 −46.8	−153 −32.0	0.9724	66.6	327.0	100.4	3:11.1
2:45	−8 −11.3	−155 −2.9	−8 −59.9	−154 −37.0	−8 −35.6	−154 −50.0	0.9721	64.5	323.1	101.2	3:10.4
2:50	−9 −1.7	−156 −23.8	−9 −50.8	−155 −57.6	−9 −26.2	−156 −10.7	0.9718	62.2	319.7	102.2	3:9.7
2:55	−9 −53.9	−157 −48.1	−10 −43.8	−157 −21.8	−10 −18.8	−157 −35.0	0.9715	59.8	316.8	103.3	3:8.9
3:00	−10 −48.3	−159 −16.6	−11 −39.0	−158 −50.3	−11 −13.6	−159 −3.4	0.9711	57.3	314.2	104.6	3:8.0
3:05	−11 −44.9	−160 −50.1	−12 −36.5	−160 −23.8	−12 −10.7	−160 −37.0	0.9707	54.6	311.9	106.1	3:7.1
3:10	−12 −44.0	−162 −29.9	−13 −36.8	−162 −3.7	−13 −10.3	−162 −16.8	0.9702	51.8	309.8	107.8	3:6.1
3:15	−13 −45.9	−164 −17.1	−14 −40.1	−163 −51.2	−14 −12.9	−164 −4.2	0.9696	48.9	307.8	109.8	3:5.1
3:20	−14 −51.1	−166 −13.8	−15 −46.8	−165 −48.2	−15 −18.9	−166 −1.0	0.9690	45.7	305.9	112.0	3:4.0
3:25	−16 0.00	−168 −22.2	−16 −57.6	−167 −57.3	−16 −28.8	−168 −9.7	0.9683	42.4	304.1	114.5	3:2.8
3:30	−17 −13.5	−170 −45.8	−18 −13.4	−170 −22.0	−17 −43.4	−170 −33.9	0.9676	38.8	302.3	117.3	3:1.6
3:35	−18 −32.8	−173 −29.8	−19 −35.4	−173 −7.6	−19 −4.0	−173 −18.6	0.9667	34.9	300.4	120.7	3:0.3
3:40	−19 −59.7	−176 −42.3	−21 −5.6	−176 −22.9	−20 −32.5	−176 −32.5	0.9656	30.5	298.4	124.6	2:58.8
3:45	−21 −37.5	179 21.2	−22 −48.0	179 35.6	−22 −12.6	179 28.6	0.9644	25.4	296.1	129.4	2:57.1
3:50	−23 −34.0	174 5.9	−24 −51.4	174 9.2	−24 −12.5	174 8.0	0.9628	19.0	293.2	135.7	2:55.1
3:55	−26 −18.5	165 13.3	−27 −54.4	164 25.6	−27 −5.5	164 52.3	0.9602	9.2	288.4	146.1	2:52.2
Limits	−28 −45.0	155 3.6	−30 −8.6	155 19.7	−29 −26.8	155 11.7	0.9577	0.0	283.7	156.2	2:49.7

Eclipse predictions and map courtesy of F. Espenak – NASA/GSFC.

Table 5.8. *Annular solar eclipse of Tuesday, February 16, 1999*

Saros 140 $\Delta T = 62.9$ sec

Universal Time	Northern limit		Southern limit		Center line		Diameter ratio	Sun alt.	Sun az.	Path width	Duration annularity
	Latitude	Longitude	Latitude	Longitude	Latitude	Longitude					
Limits	−41 −7.7	−8 −7.6	−41 −54.8	−7 −35.5	−41 −31.2	−7 −51.7	0.9774	0.0	106.8	96.3	1:18.6
5:00	−44 −20.2	−23 −49.6	−44 −40.3	−21 −26.5	−44 −30.7	−22 −41.7	0.9808	11.7	96.1	80.6	1:13.1
5:05	−45 −50.2	−33 −49.6	−46 −19.5	−32 −37.0	−46 −4.8	−33 −14.1	0.9831	20.0	87.4	69.7	1:8.6
5:10	−46 −34.8	−40 −54.2	−47 −4.9	−40 −6.1	−46 −49.8	−40 −30.5	0.9847	25.8	80.6	62.5	1:5.2
5:15	−46 −58.1	−46 −41.0	−47 −27.3	−46 −7.0	−47 −12.7	−46 −24.3	0.9860	30.4	74.6	57.0	1:2.2
5:20	−47 −7.8	−51 −40.4	−47 −35.7	−51 −16.0	−47 −21.7	−51 −28.4	0.9870	34.5	69.0	52.4	0:59.6
5:25	−47 −7.5	−56 −6.7	−47 −34.1	−55 −49.2	−47 −20.8	−55 −58.0	0.9878	38.0	63.7	48.6	0:57.2
5:30	−46 −59.7	−60 −7.8	−47 −24.8	−59 −55.6	−47 −12.2	−60 −1.7	0.9886	41.2	58.4	45.3	0:54.9
5:35	−46 −45.6	−63 −48.8	−47 −9.4	−63 −40.7	−46 −57.5	−63 −44.8	0.9893	44.2	53.2	42.5	0:52.9
5:40	−46 −26.4	−67 −13.3	−46 −48.9	−67 −8.4	−46 −37.6	−67 −10.9	0.9899	46.9	48.0	40.0	0:51.0
5:45	−46 −2.8	−70 −23.7	−46 −24.0	−70 −21.5	−46 −13.3	−70 −22.6	0.9904	49.3	42.7	37.9	0:49.2
5:50	−45 −35.3	−73 −22.2	−45 −55.4	−73 −22.1	−45 −45.3	−73 −22.1	0.9908	51.6	37.2	36.0	0:47.6
5:55	−45 −4.5	−76 −10.1	−45 −23.4	−76 −11.8	−45 −13.9	−76 −11.0	0.9912	53.6	31.7	34.4	0:46.1
6:00	−44 −30.6	−78 −48.9	−44 −48.6	−78 −52.1	−44 −39.6	−78 −50.5	0.9916	55.5	25.9	33.1	0:44.8
6:05	−43 −54.0	−81 −19.7	−44 −11.1	−81 −24.1	−44 −2.5	−81 −21.9	0.9919	57.1	19.9	31.9	0:43.6
6:10	−43 −14.9	−83 −43.3	−43 −31.3	−83 −48.7	−43 −23.1	−83 −46.0	0.9921	58.5	13.6	30.9	0:42.5
6:15	−42 −33.5	−86 −0.7	−42 −49.3	−86 −7.0	−42 −41.4	−86 −3.9	0.9923	59.6	7.2	30.1	0:41.6
6:20	−41 −50.1	−88 −12.6	−42 −5.2	−88 −19.7	−41 −57.6	−88 −16.1	0.9925	60.5	0.5	29.5	0:40.9
6:25	−41 −4.6	−90 −19.7	−41 −19.2	−90 −27.5	−41 −11.9	−90 −23.6	0.9926	61.1	353.7	29.1	0:40.3
6:30	−40 −17.3	−92 −22.7	−40 −31.5	−92 −31.1	−40 −24.3	−92 −26.9	0.9927	61.5	346.8	28.9	0:39.8
6:35	−39 −28.1	−94 −22.1	−39 −42.0	−94 −31.2	−39 −35.0	−94 −26.7	0.9928	61.6	339.8	28.8	0:39.6
6:40	−38 −37.2	−96 −18.7	−38 −50.9	−96 −28.4	−38 −44.0	−96 −23.5	0.9928	61.4	333.0	28.9	0:39.4
6:45	−37 −44.6	−98 −12.9	−37 −58.1	−98 −23.3	−37 −51.3	−98 −18.1	0.9927	60.9	326.4	29.2	0:39.5
6:50	−36 −50.3	−100 −5.4	−37 −3.7	−100 −16.5	−36 −57.0	−100 −11.0	0.9927	60.1	320.0	29.6	0:39.7
6:55	−35 −54.3	−101 −56.9	−36 −7.6	−102 −8.6	−36 −1.0	−102 −2.7	0.9925	59.1	313.9	30.3	0:40.0
7:00	−34 −56.6	−103 −47.8	−35 −10.0	−104 −0.4	−35 −3.3	−103 −54.1	0.9924	57.8	308.3	31.1	0:40.5
7:05	−33 −57.1	−105 −39.0	−34 −10.6	−105 −52.5	−34 −3.8	−105 −45.7	0.9922	56.3	303.0	32.1	0:41.2
7:10	−32 −55.8	−107 −31.1	−33 −9.4	−107 −45.6	−33 −2.6	−107 −38.3	0.9919	54.6	298.0	33.4	0:42.0
7:15	−31 −52.5	−109 −25.0	−32 −6.3	−109 −40.7	−31 −59.3	−109 −32.8	0.9916	52.6	293.5	34.8	0:43.0
7:20	−30 −47.0	−111 −21.7	−31 −1.1	−111 −38.7	−30 −54.0	−111 −30.2	0.9913	50.5	289.3	36.6	0:44.2
7:25	−29 −39.2	−113 −22.4	−29 −53.7	−113 −40.9	−29 −46.4	−113 −31.6	0.9909	48.1	285.4	38.6	0:45.5
7:30	−28 −28.8	−115 −28.4	−28 −43.7	−115 −48.7	−28 −36.2	−115 −38.5	0.9904	45.5	281.8	40.9	0:47.0
7:35	−27 −15.3	−117 −41.6	−27 −30.7	−118 −4.1	−27 −23.0	−117 −52.8	0.9898	42.7	278.5	43.5	0:48.7
7:40	−25 −58.3	−120 −4.4	−26 −14.2	−120 −29.6	−26 −6.2	−120 −17.0	0.9892	39.6	275.4	46.5	0:50.6
7:45	−24 −36.9	−122 −40.4	−24 −53.4	−123 −9.0	−24 −45.1	−122 −54.6	0.9884	36.2	272.5	49.9	0:52.8
7:50	−23 −9.9	−125 −34.6	−23 −27.0	−126 −7.9	−23 −18.4	−125 −51.1	0.9876	32.4	269.7	53.9	0:55.1
7:55	−21 −35.3	−128 −55.8	−21 −52.8	−129 −35.7	−21 −44.0	−129 −15.6	0.9865	28.1	267.1	58.6	0:57.9
8:00	−19 −49.2	−133 −0.7	−20 −6.7	−133 −51.6	−19 −57.9	−133 −25.8	0.9852	22.9	264.5	64.4	1:1.0
8:05	−17 −41.7	−138 −31.1	−17 −57.1	−139 −46.5	−17 −49.4	−139 −7.8	0.9834	16.1	261.8	72.1	1:5.1
Limits	−13 −13.2	−153 −54.8	−13 −58.3	−154 −17.7	−13 −35.7	−154 −6.2	0.9788	0.0	257.2	90.1	1:13.7

Eclipse predictions and map courtesy of F. Espenak – NASA/GSFC.

Table 5.9. *Total solar eclipse of Wednesday, August 11, 1999*

Saros 145 $\Delta T = 63.3$ sec

Universal Time	Northern limit Latitude	Northern limit Longitude	Southern limit Latitude	Southern limit Longitude	Center line Latitude	Center line Longitude	Diameter ratio	Sun alt.	Sun az.	Path width	Duration totality
Limits	41 16.4	65 17.3	40 47.7	64 54.3	41 2.0	65 5.7	1.0143	0.0	69.5	60.8	0:46.5
9:35	46 7.3	46 42.8	45 47.8	44 48.2	45 58.0	45 44.5	1.0189	15.6	83.9	78.8	1:9.1
9:40	47 57.1	37 47.7	47 24.9	36 18.8	47 41.2	37 2.6	1.0208	22.4	91.6	85.8	1:20.4
9:45	49 4.0	30 58.5	48 24.4	29 43.9	48 44.4	30 20.6	1.0222	27.6	98.3	90.5	1:29.3
9:50	49 47.9	25 11.3	49 3.0	24 7.8	49 25.6	24 39.1	1.0233	31.9	104.4	94.1	1:37.0
9:55	50 16.3	20 3.3	49 27.4	19 9.6	49 51.9	19 36.0	1.0242	35.7	110.3	97.1	1:43.8
10:00	50 32.9	15 23.5	49 40.9	14 38.8	50 6.9	15 0.7	1.0250	39.1	116.1	99.5	1:49.9
10:05	50 40.1	11 5.4	49 45.6	10 29.3	50 12.9	10 47.0	1.0257	42.1	121.8	101.6	1:55.4
10:10	50 39.4	7 5.0	49 43.1	6 37.0	50 11.3	6 50.7	1.0263	44.8	127.6	103.3	2:0.3
10:15	50 32.1	3 19.7	49 34.3	2 59.4	50 3.2	3 9.3	1.0268	47.3	133.4	104.9	2:4.7
10:20	50 18.8	0 −12.7	49 20.1	0 −25.6	49 49.4	0 −19.4	1.0272	49.6	139.3	106.2	2:8.6
10:25	50 0.4	−3 −33.6	49 1.0	−3 −39.6	49 30.7	−3 −36.8	1.0275	51.6	145.4	107.4	2:12.1
10:30	49 37.3	−6 −44.3	48 37.6	−6 −43.8	49 7.4	−6 −44.1	1.0278	53.4	151.6	108.4	2:15.0
10:35	49 9.9	−9 −45.8	48 10.2	−9 −39.2	48 40.1	−9 −42.6	1.0281	55.0	158.0	109.3	2:17.5
10:40	48 38.8	−12 −39.1	47 39.3	−12 −26.8	48 9.0	−12 −33.0	1.0283	56.4	164.6	110.1	2:19.5
10:45	48 4.0	−15 −25.0	47 5.0	−15 −7.4	47 34.5	−15 −16.2	1.0284	57.5	171.4	110.7	2:21.1
10:50	47 26.1	−18 −4.1	46 27.7	−17 −41.7	46 56.9	−17 −52.9	1.0285	58.4	178.3	111.3	2:22.2
10:55	46 45.0	−20 −37.3	45 47.5	−20 −10.5	46 16.2	−20 −23.8	1.0286	59.0	185.3	111.7	2:22.8
11:00	46 1.1	−23 −5.2	45 4.5	−22 −34.3	45 32.8	−22 −49.7	1.0286	59.3	192.4	112.1	2:23.0
11:05	45 14.4	−25 −28.5	44 18.9	−24 −53.9	44 46.7	−25 −11.1	1.0286	59.3	199.4	112.3	2:22.7
11:10	44 25.1	−27 −47.7	43 30.8	−27 −9.8	43 58.0	−27 −28.6	1.0285	59.1	206.3	112.5	2:22.0
11:15	43 33.2	−30 −3.7	42 40.3	−29 −22.7	43 6.8	−29 −43.1	1.0284	58.6	213.1	112.5	2:20.9
11:20	42 38.8	−32 −17.0	41 47.3	−31 −33.3	42 13.1	−31 −55.0	1.0282	57.8	219.6	112.5	2:19.3
11:25	41 41.9	−34 −28.4	40 52.0	−33 −42.2	41 17.0	−34 −5.1	1.0280	56.8	225.7	112.2	2:17.3
11:30	40 42.5	−36 −38.5	39 54.2	−35 −50.2	40 18.4	−36 −14.2	1.0278	55.5	231.6	111.9	2:14.9
11:35	39 40.5	−38 −48.3	38 53.9	−37 −57.9	39 17.3	−38 −22.9	1.0274	53.9	237.1	111.4	2:12.0
11:40	38 35.8	−40 −58.5	37 51.0	−40 −6.4	38 13.5	−40 −32.3	1.0271	52.2	242.2	110.7	2:8.8
11:45	37 28.3	−43 −10.3	36 45.4	−42 −16.5	37 6.9	−42 −43.2	1.0267	50.2	247.0	109.8	2:5.1
11:50	36 17.7	−45 −24.7	35 36.7	−44 −29.4	35 57.3	−44 −56.9	1.0262	48.0	251.5	108.6	2:1.0
11:55	35 3.7	−47 −43.3	34 24.8	−46 −46.6	34 44.4	−47 −14.8	1.0256	45.6	255.6	107.2	1:56.5
12:00	33 45.9	−50 −7.9	33 9.3	−49 −9.8	33 27.7	−49 −38.7	1.0250	42.9	259.5	105.4	1:51.5
12:05	32 23.6	−52 −40.9	31 49.4	−51 −41.3	32 6.6	−52 −10.9	1.0243	39.9	263.2	103.1	1:46.0
12:10	30 55.9	−55 −25.6	30 24.4	−54 −24.4	30 40.3	−54 −54.8	1.0235	36.7	266.6	100.3	1:40.1
12:15	29 21.4	−58 −27.1	28 52.9	−57 −23.9	29 7.3	−57 −55.3	1.0225	33.0	269.9	96.9	1:33.5
12:20	27 37.8	−61 −53.7	27 12.7	−60 −47.7	27 25.4	−61 −20.5	1.0214	28.9	273.1	92.5	1:26.1
12:25	25 40.3	−66 −1.1	25 19.8	−64 −50.3	25 30.3	−65 −25.4	1.0201	24.0	276.2	86.9	1:17.7
12:30	23 18.2	−71 −27.3	23 4.5	−70 −6.0	23 11.6	−70 −46.3	1.0182	17.7	279.5	79.1	1:7.5
12:35	19 32.9	−81 −37.0	19 44.3	−79 −9.3	19 40.0	−80 −19.1	1.0151	7.1	283.8	64.9	0:51.7
Limits	17 46.9	−87 −24.9	17 20.1	−87 −9.0	17 33.5	−87 −16.9	1.0129	0.0	286.1	55.2	0:42.2

Eclipse predictions and map courtesy of F. Espenak – NASA/GSFC.

each table is the type of eclipse (annular or total) and date, followed by the saros series and the extrapolated value of ΔT which was used in the calculations. The first column of each table gives the Universal Time (UT) for the data which follow. Depending on the duration of the central eclipse, data are listed at one, two, four or six minute intervals. The next six columns define the geodetic coordinates of the northern and southern limits as well as the center line. The latitude and longitude of each point are given in degrees and minutes to the nearest tenth of a minute. Negative latitudes are south of the Equator and negative longitudes are east of the Greenwich Meridian.

The column identified as "Diameter ratio," is the ratio of the topocentric apparent diameters of the moon and the sun. For total eclipses, the ratio is always greater than or equal to 1.000. For annular eclipses, the ratio is less than 1.000 and is identical to the eclipse magnitude at maximum eclipse. Eclipse magnitude is defined as the fraction of the sun's diameter obscured by the moon. The next two columns, "Sun alt," and "Sun az," give the sun's altitude and azimuth at maximum eclipse as seen by an observer on the center line. The eleventh column lists the width of the path of totality or annularity in kilometers. Finally, the duration of the total or annular phase is given in minutes and seconds.

All path characteristics are calculated for sea level and the effects of refraction have been ignored.

Partial eclipses

In addition to the total and annular eclipses during 1994–2000, the following eclipses never become more than partial, because the tip of the moon's shadow cone passes north or south of the earth.

Partial solar eclipse of April 17, 1996 These partial phases, reaching a peak magnitude of 88% at 22:37:10 UT, are visible only from New Zealand and the north coast of Antarctica.

Partial solar eclipse of October 12, 1996 These partial phases, reaching a peak magnitude of 76% at 14:02:00 UT, will be widely visible in Europe and North Africa, with the southern limit passing through the Sahara. A small amount of sun will be eclipsed as seen from northernmost Maine, most of New Brunswick, northeastern Nova Scotia, Newfoundland and northern Canada from Hudson Bay on eastward as well as Greenland and Iceland. The eastern limit will pass from near St Petersburg south through the Caspian Sea and Saudi Arabia.

Partial solar eclipse of September 2, 1997 These partial phases, reaching a peak magnitude of 90% at 0:03:43 UT, will be visible only from Australia, New Zealand, and Antarctica.

CHAPTER SIX

THE TOTAL SOLAR ECLIPSE OF AUGUST 11, 1999, IN EUROPE

People in Europe will have an excellent opportunity to view a total solar eclipse on August 11, 1999. The eclipse is a relatively short one, with totality lasting less than 2½ minutes, but totality will be visible from several major cities, weather permitting.

The path of totality begins in the North Atlantic off Nova Scotia but first touches land in Cornwall. Then it crosses the Channel, passes to the north of Paris, then through Stuttgart, Munich, Salzburg, Graz, and Bucharest, and continues ultimately through the Middle East to India.

Site selection

Clear weather in Europe is far from a certainty; observers from other parts of the world should be aware of the substantial risk of being "clouded out." Central Europe is the worst afflicted; the weather outlook is slightly better at both ends of the path. Western England enjoys a slight advantage over northern France. The best sites in Europe, however, are on the Romanian and Bulgarian coast, where the weather is almost as clear as in the American southwest.

The sun is at least 45° above the horizon at the time of the eclipse throughout Europe and as far east as Iraq.

Time zones

Throughout Europe, summer time (daylight saving time) will be in effect at the time of the eclipse. The time zones involved are:

BST (British Summer Time)	UT + 1
MESZ (Mitteleuropasommerzeit)	UT + 2
OESZ (Osteuropasommerzeit)	UT + 3

Because changes are possible, all time zone information should be confirmed locally shortly before the eclipse.

Maps and tables

Maps 6.1–6.3 show the path of totality in Europe, with the time of mid-eclipse at 10-minute intervals. (See Fig. 5.1, page 68, for a sketch of the complete path.) Table 6.1 gives the entire path in numerical forms, listing latitude and longitude as functions of time. Table 6.2, interpolated from Table 6.1, gives latitude as a function of longitude for easier plotting on maps. Table 5.9 (page 95) provides some additional information, and a map of the earth showing where the total and partial phases will be visible appears on page 85.

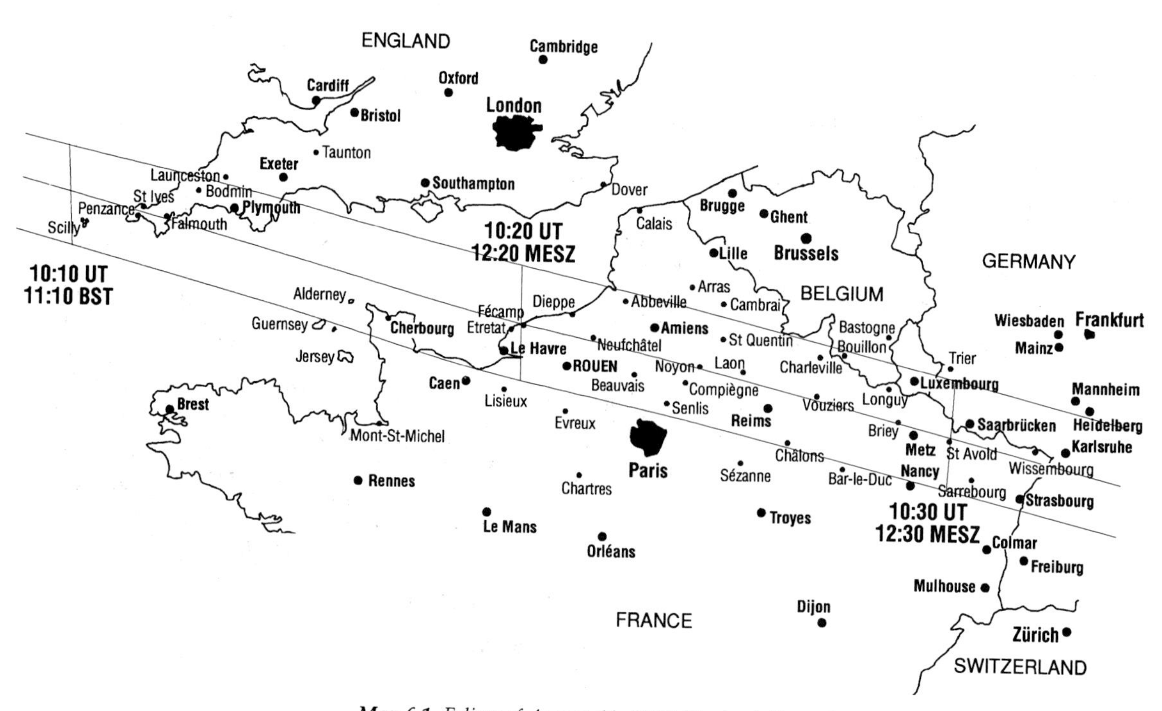

***Map 6.1.** Eclipse of August 11, 1999 (England, France).*

Map 6.2. *Eclipse of August 11, 1999 (Germany, Austria).*

Map 6.3. *Eclipse of August 11, 1999 (Hungary, Romania, Bulgaria).*

Table 6.1. *Path of totality, Wednesday, August 11, 1999. From* Circular No. 170, *US Naval Observatory.*

UT		Northern limit				Central line				Southern limit				Central line		
		Latitude		Longitude		Latitude		Longitude		Latitude		Longitude		Duration		Alt.
h	m	°	′	°	′	°	′	°	′	°	′	°	′	m	s	°
Limit		+41	16.9	−65	17.7	+41	01.7	−65	05.5	+40	46.5	−64	53.5	0	49.2	
9	35	+46	07.4	−46	45.6	+45	57.6	−45	44.6	+45	46.9	−44	45.8	1	12.2	16
9	40	+47	57.3	−37	49.6	+47	40.9	−37	02.7	+47	23.8	−36	17.2	1	23.6	22
9	45	+49	04.4	−31	00.1	+48	44.0	−30	20.7	+48	23.3	−29	42.6	1	32.7	28
9	50	+49	48.4	−25	12.6	+49	25.3	−24	39.2	+49	01.9	−24	06.9	1	40.5	32
9	55	+50	16.8	−20	04.5	+49	51.6	−19	36.2	+49	26.2	−19	08.9	1	47.5	36
10	00	+50	33.4	−15	24.4	+50	06.6	−15	00.9	+49	39.6	−14	38.3	1	53.6	39
10	05	+50	40.7	−11	06.2	+50	12.6	−10	47.2	+49	44.4	−10	28.9	1	59.2	42
10	10	+50	40.0	−7	05.7	+50	11.0	−6	51.0	+49	41.8	−6	36.8	2	04.2	45
10	15	+50	32.7	−3	20.3	+50	02.9	−3	09.5	+49	33.0	−2	59.4	2	08.7	47
10	20	+50	19.4	−0	12.2	+49	49.1	−0	19.1	+49	18.8	+0	25.5	2	12.7	50
10	25	+50	01.0	+3	33.2	+49	30.4	+3	36.5	+48	59.7	+3	39.4	2	16.1	52
10	30	+49	37.9	+6	44.0	+49	07.1	+6	43.8	+48	36.3	+6	43.5	2	19.1	53
10	35	+49	10.5	+9	45.6	+48	39.8	+9	42.3	+48	09.0	+9	38.8	2	21.7	55
10	40	+48	39.4	+12	38.9	+48	08.7	+12	32.6	+47	38.1	+12	26.3	2	23.7	56
10	45	+48	04.6	+15	24.9	+47	34.2	+15	15.8	+47	03.8	+15	06.8	2	25.3	58
10	50	+47	26.7	+18	04.1	+46	56.6	+17	52.5	+46	26.5	+17	41.0	2	26.4	58
10	55	+46	45.6	+20	37.3	+46	16.0	+20	23.4	+45	46.3	+20	09.7	2	27.1	59
11	00	+46	01.6	+23	05.3	+45	32.5	+22	49.2	+45	03.4	+22	33.4	2	27.2	59
11	05	+45	15.0	+25	28.6	+44	46.4	+25	10.6	+44	17.8	+24	52.9	2	27.0	59
11	10	+44	25.6	+27	47.9	+43	57.7	+27	28.2	+43	29.8	+27	08.8	2	26.3	59
11	15	+43	33.7	+30	03.9	+43	06.5	+29	42.6	+42	39.3	+29	21.7	2	25.1	59
11	20	+42	39.3	+32	17.2	+42	12.9	+31	54.5	+41	46.3	+31	32.2	2	23.5	58
11	25	+41	42.4	+34	28.6	+41	16.8	+34	04.7	+40	51.0	+33	41.1	2	21.5	57
11	30	+40	43.0	+36	38.8	+40	18.2	+36	13.7	+39	53.2	+35	49.0	2	19.0	55
11	35	+39	41.0	+38	48.6	+39	17.0	+38	22.4	+38	53.0	+37	56.7	2	16.1	54
11	40	+38	36.3	+40	58.9	+38	13.3	+40	31.8	+37	50.1	+40	05.1	2	12.8	52
11	45	+37	28.8	+43	10.6	+37	06.7	+42	42.7	+36	44.5	+42	15.1	2	09.1	50
11	50	+36	18.2	+45	25.1	+35	57.1	+44	56.3	+35	35.9	+44	28.0	2	04.9	48
11	55	+35	04.2	+47	43.7	+34	44.2	+47	14.2	+34	24.0	+46	45.1	2	00.4	46
12	00	+33	46.3	+50	08.3	+33	27.5	+49	38.0	+33	08.5	+49	08.2	1	55.3	43
12	05	+32	24.1	+52	41.3	+32	06.5	+52	10.2	+31	48.6	+51	39.6	1	49.8	40
12	10	+30	56.4	+55	26.0	+30	40.2	+54	54.1	+30	23.7	+54	22.6	1	43.7	37
12	15	+29	21.9	+58	27.6	+29	07.2	+57	54.6	+28	52.2	+57	22.0	1	37.0	33
12	20	+27	38.2	+61	54.3	+27	25.3	+61	19.7	+27	12.1	+60	45.6	1	29.6	29
12	25	+25	40.7	+66	01.6	+25	30.2	+65	24.5	+25	19.3	+64	47.9	1	21.0	24
12	30	+23	18.5	+71	28.0	+23	11.7	+70	45.0	+23	04.3	+70	03.0	1	10.6	18
12	35	+19	33.0	+81	38.4	+19	40.5	+80	16.2	+19	44.9	+79	03.2	0	54.7	7
Limit		+17	47.5	+87	25.1	+17	33.2	+87	16.6	+17	18.9	+87	08.2	0	45.0	

Table 6.2. *Path of totality, Wednesday, August 11, 1999. Interpolated table of latitude against longitude, for plotting on your own maps (European sites only).*

Longitude	Latitude			Longitude	Latitude		
	Northern limit	Center of path	Southern limit		Northern limit	Center of path	Southern limit
W 10.0	N 50.686	50.213	49.740	E 15.0	N 48.169	47.629	47.089
W 9.0	N 50.686	50.210	49.734	E 16.0	N 47.944	47.401	46.857
W 8.0	N 50.679	50.201	49.722	E 17.0	N 47.709	47.162	46.614
W 7.0	N 50.665	50.186	49.704	E 18.0	N 47.462	46.911	46.360
W 6.0	N 50.646	50.166	49.684	E 19.0	N 47.204	46.650	46.094
W 5.0	N 50.618	50.138	49.656	E 20.0	N 46.934	46.377	45.818
W 4.0	N 50.579	50.097	49.614	E 21.0	N 46.652	46.091	45.529
W 3.0	N 50.525	50.038	49.551	E 22.0	N 46.359	45.794	45.229
W 2.0	N 50.459	49.960	49.463	E 23.0	N 46.054	45.485	44.917
W 1.0	N 50.384	49.875	49.366	E 24.0	N 45.738	45.165	44.592
0.0	N 50.308	49.793	49.278	E 25.0	N 45.411	44.833	44.256
E 1.0	N 50.233	49.719	49.203	E 26.0	N 45.070	44.489	43.909
E 2.0	N 50.155	49.645	49.132	E 27.0	N 44.717	44.133	43.550
E 3.0	N 50.070	49.564	49.055	E 28.0	N 44.352	43.765	43.180
E 4.0	N 49.971	49.466	48.960	E 29.0	N 43.976	43.385	42.797
E 5.0	N 49.857	49.350	48.842	E 30.0	N 43.587	42.994	42.401
E 6.0	N 49.731	49.220	48.709	E 31.0	N 43.187	42.591	41.994
E 7.0	N 49.595	49.080	48.565	E 32.0	N 42.775	42.177	41.577
E 8.0	N 49.450	48.933	48.415	E 33.0	N 42.352	41.751	41.149
E 9.0	N 49.297	48.778	48.257	E 34.0	N 41.918	41.315	40.711
E 10.0	N 49.135	48.614	48.090	E 35.0	N 41.472	40.867	40.261
E 11.0	N 48.963	48.438	47.911	E 36.0	N 41.017	40.409	39.802
E 12.0	N 48.781	48.251	47.721	E 37.0	N 40.551	39.941	39.334
E 13.0	N 48.588	48.054	47.521	E 38.0	N 40.075	39.464	38.857
E 14.0	N 48.383	47.847	47.310	E 39.0	N 39.591	38.978	38.371

Note: Tables 6.1 and 6.2 were computed independently from Table 5.9, and so differ slightly.

CHAPTER SEVEN

VIDEOTAPING SOLAR ECLIPSES

The ubiquitous camcorder has found its way to eclipse sites. Camcorders are easy to set up, and the zoom lenses make it easy to fill the screen with the solar corona. Still, solar filters are necessary during partial phases, and the timing of when to remove and reinstall them can make the difference between a dramatic diamond ring and a fizzle.

I (JMP) have used camcorders to photograph two eclipses; my knowledge is supplemented with the camcorder experience of Fred Espenak and Marjorie Nicol. Both flew with me on a Scientific Expeditions, Inc., airplane to the 1990 eclipse in Finland. Dr. Nicol joined me on a plane out of Cape Town to view the 1992 eclipse over the ocean southwest of South Africa. Mr. Espenak collaborated with me on a videotape instruction article for *Sky & Telescope* magazine prior to the 1991 eclipse.

Video cameras have the major advantages of ease of use, wide dynamic range, and accessible zoom lenses. Ordinary 35-mm still cameras give much sharper images but do not provide the impact of change that camcorders with their moving images and their sound tracks do. As for motion-picture cameras, Super-8 home-movie cameras are all but extinct, following this camcorder revolution. Their film size was small, anyway, so the resolution (amount of detail detectable) wasn't that much better than that of camcorders. Sixteen-mm motion-picture cameras do give much higher resolution images than camcorders, but film is so expensive, not to mention the price of camera rental (let alone purchase), that these cameras are seldom used by amateurs. My own 16-mm time-lapse eclipse film, made in clear skies in collaboration with my student Stuart Vogel at the 1973 eclipse with the support of the National Geographic Society, may thus remain the best eclipse film ever made.

In the next few years, we will see high-definition television (HDTV) eclipse images, but that time has not yet come. The sole exception is an HDTV tape of the July 11, 1991, eclipse made at the top of Mauna Kea in Hawaii by a team from Japanese television. The tape is, unfortunately, of a standard not readily playable in the United States and was made in poor observing conditions.

What videos can we make now, with current equipment? Most camcorders are VHS. Images made with these camcorders can be excellent for viewing on a television. The sharpness and resolution is not adequate, though, for appearing in print or for making a print to hang on the wall. Perhaps as people get used to the somewhat lower resolution compared with film provided by the new Kodak Photo-CD, these VHS images will be more satisfactory.

An improvement on VHS, called S-VHS (for Super) is readily available, though the prices for both cameras and video-cassette recorders (VCRs) are higher than for ordinary VHS. The improvement of S-VHS over VHS is noticeable. It is also possible to get Hi-Q stereophonic sound, though this difference is not so important for eclipse work. Still, the ambiance of the eclipse location – often punctuated with cheering at the moment of totality – is nice to record.

At present, Video-8 recorders are very popular. Their small size and light weight overcome for many people the fact that the 8-mm tapes (about the size of audio cassette tapes) cannot be played on ordinary VHS VCRs. Playback units are available, but are little used. It is easy to play back from the 8-mm recorder directly to your television, or to rerecord onto a VHS VCR to make a copy that can be easily played later. (Small TVs that can play back the picture from the camcorder are available for only about US$150; they often run on batteries and so also make nice color monitors for the camcorders.)

An improved version of Video-8 called Hi-8 is now widely available. The price difference between Hi-8 and Video-8 is not great, and Hi-8 tapes are at least as good as S-VHS tapes. Hi-8 camcorders with wonderfully powerful zoom lenses are widely available for under US$1,000. New features now being introduced include electronic stabilization, so on future airplane trips to view eclipses, the images can be steadier. Small VCRs that record from the air or from camera input onto Hi-8 and also play back Hi-8 tapes in color are also available for under US$1,000.

On the 1990 airborne expedition, three of us had 8-mm camcorders, and we were all pleased with the results. The two of us on the 1992 flight still had our 8-mm camcorders, and were again pleased. An interesting bonus is that as soon as the eclipse is over, one can merely plug in wires (''patch cords'') from one camcorder to the other and make copies of the tapes. Thus even if one of us had lost a tape on the way home, a high-quality copy already existed. This safety factor is very reassuring.

Once upon a time, television cameras were very delicate. Indeed, an Apollo 12 astronaut pointed his TV camera at the sun for an instant by mistake, and the camera was permanently out of commission. Times have changed. Most of today's cameras now use charge-coupled devices (CCDs).

These CCDs are light-sensitive silicon chips that make electrons when light hits them. A computer guides in reading out the number of chips in each picture element (pixel). Since the mode in which the readout takes place involves sliding the electrons over one pixel at a time, coupling the charge between adjacent pixels, they got their name of "charge coupled." A great advantage of CCDs for eclipse work is that they can accept a brief exposure to direct sunlight without being harmed.

Most astronomy has now converted from film to CCDs as the detector. Since CCDs record about 50% of the photons of light that hit them, compared with about 2% for film, the electronic detectors of today are about 25 times more sensitive than the films of the past. Thus exposures that once took 25 minutes now can be made in about one minute. Astronomy is benefiting from CCDs across the board. But you can also note the availability of CCDs in consumer electronics. Just look at the ads for TV camcorders in your local newspaper.

Total eclipses appear nicely on screen either as wide-angle views or as closeups. The wide-angle views can be taken with the camcorder lens on its wide or medium settings. For closeup views, the lenses widely available on today's camcorders do nicely on maximum zoom. The size of the sun's image on the screen depends on the size of the CCD detector (some of which are ½-inch across and others of which are ⅔-inch across) and on the maximum focal length of the lens (Table 7.1).

The sun (and the moon) are about ½° across. But remember that at an eclipse, you want to photograph the corona, which appears for one or two times the radius of the sun to either side. So you want your closeup image to show at least 2.5°, and actually perhaps twice that to allow for space all around. You should also realize that the sun moves across the sky at a rate that is quite noticeable at high magnification if your camcorder is on a tripod. (Of course, you could use a mount that tracks at the solar rate – or even one that tracks at the normal rate of stars, the sidereal rate, which differs by only 4 minutes a day from the solar rate – but such mounts aren't often taken to eclipses.) The sun and moon move 15° per hour, which means they move 1° – twice the diameter of the sun – during the 4 minutes that is typical of totality at an eclipse. Thus if your camera is not arranged so that it will move during totality, you must give the solar image room on the screen to move.

I have thus found that the standard maximum zoom on a telephoto gives a very pleasing image of totality. The 80-mm maximum focal length on the 10:1 zoom on my Canon Hi-8 camcorder, which has a ½-inch CCD, gives an image of the sun's disk that is 1.3 inches (32 mm) across on a 13-inch-diagonal television screen. Allowing for corona on both sides, the image is 4 inches (100 mm) across and high, which fits nicely on the 8-inch-high (20-cm) screen that is known as a 13-inch TV. This size gives you a bit of room for the sun to travel during totality, and for pleasing dark space around the image.

Table 7.1. *Image sizes for camcorders*
On a 13-inch-diagonal (8-inch-high) television

Camcorder focal length (mm)	Size of sun (½-inch CCD)		Size of sun (⅔-inch CCD)	
	(mm)	(ins)	(mm)	(ins)
50	20	0.8	17	0.7
60	24	0.9	20	0.8
70	28	1.1	24	0.9
80	32	1.3	27	1.1
90	36	1.4	30	1.2
100	40	1.6	34	1.3
150	60	2.4	51	2.0
200	80	3.2	68	2.7
250	100	3.9	85	3.3
300	120	4.7	102	4.0
350	140	5.5	119	4.7
400	160	6.3	136	5.4
450	180	7.1	153	6.0
500	200	7.9	170	6.7

For a total eclipse, multiply by 3 to include 1-solar-diameter of corona all around.

I must add that my colleague Fred Espenak prefers to add a converter lens that enlarges the image another factor of two or more. Such lenses screw onto the front of your camcorder's lens. Converters that enlarge ×2, ×5 or more, even up to ×12, are not very expensive. Then you can really fill the screen with the sun. (These converters are also useful for photographing the moon on an ordinary night.) Among the disadvantages of converters is the fact that you are adding more glass surfaces, which can degrade resolution and add to internal reflections. Further, the optical quality of many of the converters is not equal to that of the original lens. In addition, if you zoom back from maximum, the converters vignette (partially darken or cut off the edge of) the image at some point.

You may find the following conversions helpful: On a 13-inch-diagonal (8-inch high) television screen, the sun's image size is approximately equal to 0.4 times the focal length of your lens, for a ½-inch CCD. For a ⅔-inch CCD, the multiplication factor is 0.34. Example: ½-inch CCD with a 80-mm lens. Multiply 80 mm by 0.4 to get 32 mm (about 1.5 inches) in diameter on a 13-inch TV. This image size, while small, is satisfactory for a solar eclipse, though I would prefer a bit more magnification.

The conversion for those who are used to 35-mm still photography is as follows: Multiply the focal length of a camcorder with a ½-inch CCD by 5.2 to get the equivalent focal length of a lens used with a 35-mm still camera. Example: 80-mm camcorder lens, multiplied by 5.2, means that it gives about the same proportion of image on the frame as an 80 × 5.2 = 420-mm focal length lens on a normal 35-mm still camera. Such a lens is a telephoto; a "normal" lens on a still camera is often considered to be 50-mm. For a ⅔-inch CCD, the multiplication factor is 4.5.

Don't forget that in videotaping, as in most photography, a steady image is desirable. If you can, use a tripod. Certainly on the ground, tripods steady images. It turns out that on airplanes, it is useful to use a tripod as well. (On the 1990 expedition, I tried a hand-held gyroscope mount to steady the frame, and suffered during totality from a steady drift that was hard to control. The fact that the gyros were stolen out of my luggage on the way home didn't help make me feel kindly about this method.)

Don't forget that it is fun to have wide-angle images as well as closeup ones. If you have a friend and a second camcorder, that would be great. The wide-angle images show the approach of the moon's shadow and its recession after totality. The fact that places outside the zone of totality are not in the moon's full shadow means that they are relatively bright, so the shadow cone shows up in silhouette.

You can use the sound capability of the camcorder to record not only general noises around but also anything special you want to say. You can read out changes in your camera settings, for example.

I have not seen any images of shadow bands taken with camcorders. Shadow bands are of very low contrast, and so are notoriously difficult to study.

Additional comments

To make still images from the camcorder, you need merely to photograph a TV screen. I took the photographs that illustrate this chapter (Figs. 7.1–7.4) with a 35-mm camera loaded with Kodak Gold 100 film and set at 1/15 second. You must make your setting slower than 1/30 second, since the TV picture is painted in that time, and you get diagonal bars appearing in your picture from this painting process if your exposure times are too short. In the photographs here, I used an exposure of *f*/4. To find it, I took a meter reading on an average-brightness screen filled with scenery from another part of my tape. I was able to get a steady image on the screen using the "pause" control on my camcorder, though I have also had success in photographing "on the fly."

Fig. 7.1. A wide-angle view of the 1990 eclipse on Hi-8 videotape, photographed from a television screen. We see Finland through clouds below, the umbra at center, and the eclipse in the sky. We also see internal reflections in the plane windows (including a coronal view to the lower right of the main image) and camera lens. (Eloise Pasachoff)

Fig. 7.2. A view of the 1990 eclipse with a Hi-8 camera, somewhat zoomed in from the previous photo. We see multiple reflections in the plane windows. All four images in this section were taken with hand-held cameras. (Eloise Pasachoff)

Don't forget to make your solar filters' mount easy to take off and to put on. As for ordinary cameras or for viewing, you can use Mylar filters or glass filters with a chromium deposit on them. The Mylar filters available from Roger Tuthill, Inc. (address on p. 37), come both in ordinary grade and in a super-dense grade that is usually better for videotaping the sun. The blue color that comes through a Mylar filter can be made more pleasing by sandwiching a Kodak Wratten 21 filter (which can be ordered inexpensively from your local camera store) with the Mylar filter. Glass filters, which can be ordered from Thousand Oaks Optical (address on p. 37), have an orange or yellow cast because they allow slightly more red light to pass through than other colors; this color more closely matches our perceptions of the sun. When you mount your filter, look through the viewfinder to make certain there are no internal reflections between the filter and the many internal lens surfaces. If you find some, adjusting the angle of the filter often helps.

Fig. 7.3. *A zoomed-in view of the 1990 eclipse, with the 8-mm video camera (standard-8) using its standard exposure. (Marjorie Nicol)*

Fig. 7.4. *Another zoomed-in view of the 1990 eclipse, with the same camera using its high shutter speed. (Marjorie Nicol)*

You can take off your solar filter during the diamong ring before totality, and replace it a few seconds (perhaps 5 seconds) after the beginning of the final diamond ring. In this way, you can record the whole sequence of the eclipse, including partial phases. Of course, during totality, you should not have any filter at all on the camera. Also, of course, you never take off the filter during an annular eclipse.

Remember that camera batteries often last only half an hour, so you may want to put in a fresh battery about 15 minutes before totality. (If necessary, buy a spare battery weeks in advance, when you are sure you can order it in time. Or get an adapter to line current or to a car battery if you are sure such current will be available to you.) Don't forget to turn the camera back on! And don't make any changes in the last five minutes. All too often, something goes wrong, and you are scrambling at a time when you should be relaxed.

If you can check your camcorder on the moon a week or so before the eclipse, do so. Remember that the corona at an eclipse is about the same brightness as the moon, and that the sun is about the same size as the moon. I do point out that a solar eclipse occurs when the moon passes in front of the sun, meaning that the moon is in its final, narrow, almost invisible stages for the night or two before eclipse day. So don't wait until then to try to do your tests.

Annular eclipses

At an annular eclipse, the everyday surface of the sun is never entirely covered, so you must always keep the filter on. The appearance of the sun changes gradually, so if your camcorder has a slow speed, or time lapse, it would be good to use it.

Partial phases

Since the partial phases take hours, you don't want to keep your camera running throughout. (If you do, make sure your batteries are charged, or that you are using an AC adapter or a car-battery adapter.) If your camcorder has a time-lapse mode, use it. It would be nice to reduce the, say, 3 hours of partial phases to 3 minutes, so you want to take one exposure every 2 seconds instead of the normal 30 frames per second. This lapse rate would speed up the eclipse by a factor of 60.

Do note that the sun and moon move across the sky by about 15° per hour, so you would have to have your camera mounted on a tracking mount if you want to make such a time-lapse image with a telephoto setting.

If you don't have a tracking mount, or don't have a time-lapse mode, you could take a few seconds of exposure through the filters every couple of minutes, centering the image each time. In fact, if you draw a circle the size of the sun's image on a clear piece of plastic and affix it to the viewfinder, you can line up the sun more precisely. Centering the sun by eye will give you an image that appears to jump around unpleasantly, especially since it is hard to keep the same center as the sun becomes more and more of a crescent.

Additional comments

The normal exposure of a camcorder seems to give an image that is pleasing but that very much overexposes the inner corona. I prefer the images that appear with the "high-speed shutter" available in some of the more complicated (and expensive) camcorders. But if you don't have this feature, don't worry. If you have such a feature, you can decide whether you want to run through all the possibilities during totality, or if you would rather have a smooth sequence throughout totality without any changes. The choice is yours.

You can choose to imprint the date and perhaps the time on your eclipse video, using the facility available on most camcorders. You may choose, though, not to have such extraneous information appearing on the beautiful television screen image of the eclipse.

Final comments

Do make sure, shortly before totality, that your camera is really recording! Look in the viewfinder to see that it says REC, or whatever it is supposed to say. As with many things, practice makes perfect.

Most important, remember that the camera will run itself, and that you can see the videotape later. Devote your time during the eclipse to other things – either other cameras or perhaps even looking at the eclipse directly. Enjoy the eclipse!

CHAPTER EIGHT

A PERSONAL ESSAY: THE 1991 TOTAL SOLAR ECLIPSE†

"It was like a dream come true," said Julietta Fierro. The shadow of the moon had come racing across the earth's surface, making a total solar eclipse. Professor Fierro, the National Coordinator for Mexico, observed the eclipse from La Paz, Baja California.

"We got what we needed," said Donald Hall, about the scientists 14,000 feet (3,600 meters) high on Mauna Kea in Hawaii. Dr. Hall is Director of the Institute for Astronomy of the University of Hawaii. "But it was close. Twenty minutes before totality, fog was creeping up around the domes. With the cirrus overhead and the dust from Mt. Pinatubo above it, we will have to work hard to interpret the infrared data." Rarely does a major observatory fall within the long, skinny path that an eclipse traces across the earth's surface. Excitement on Mauna Kea, site of a half-dozen of the largest and most advanced telescopes on earth, had been high.

For years, both professional and amateur astronomers had been pointing toward July 11, 1991, waiting for its eclipse. Eclipses run in cycles that last 18 years, 11⅓ days, and the wait had begun moments after the corresponding eclipse in 1973. My own research, since the 1970 eclipse Donald Menzel and I described in *National Geographic* (August 1970), had included many eclipse expeditions and many views of the solar corona, the faint outer layer of the sun. For this, my 17th eclipse expedition, the 90% prediction of clear weather enabled us with great good hope to mount an expedition of two tons of equipment, two colleagues, and a dozen students. My research was sponsored by the US National Science Foundation and the Committee on Research and Exploration of the National Geographic Society. I was also to write an article about the eclipse for *National Geographic*, an article that appeared in May 1992, accompanied by photographs by the photographer Roger Ressmeyer.

† This chapter is the report of one of us, JMP, about his expedition to the 1991 eclipse.

The chance of clear weather in certain zones of Hawaii and of Baja California was the same 90%, but which to choose? Statistics may be accurate, but they don't tell you what will actually happen. Should we opt for the traditionally clear Hawaiian early morning air for four minutes of totality or should we suffer the turbulent atmosphere at high noon in Baja California to gain over two more minutes? Should we try to set up equipment in the difficult working circumstances of 14,000-foot altitude on Mauna Kea or stay closer to sea level? I chose the sea-level site in Hawaii.

Why go to eclipses? For amateur astronomers and for tourists, it is the sheer beauty of the event. For professional astronomers like me, it is the chance to study a layer of the sun that is so faint that it is usually hidden behind the blue sky. Only by making the sun visible in a sky that is dark as night can we view the corona as a whole from the earth. Special instruments on a few high mountains can observe some aspects of the corona without eclipses, but their view is limited and specialized. Some devices aboard spacecraft have made a sort of artificial eclipse, but they have had to hide not only the everyday surface of the sun but also the inner and middle corona around it, because these lower parts of the corona are too bright to allow the higher corona to be seen. And anyway, no such coronagraph is now aloft. Solar eclipses still provide the best way of taking the latest, best equipment to observe the sun's outer part, elbowing aside the blue sky. For this eclipse, the latest equipment is electronic arrays sensitive to the infrared. Such equipment has not yet been readied for space observations. We do get some information from space, though. Observations with spacecraft like the Hubble Space Telescope and nighttime telescopes are increasingly telling us about coronas around other stars, so what we learn about the sun is automatically pertinent for millions of other objects as well.

On July 11, 1991, the moon's shadow was about 200 miles (320 km) wide as it hit the earth. As the shadow moved through space and as the earth rotated, the shadow met the earth as an ellipse moving thousands of miles per hour. The shadow reached earth near Hawaii about an hour after sunrise there and then crossed over the Pacific Ocean. It next met land near noontime at the southern tip of Baja California, over 1,000 miles (1,600 km) south of the California border. The earth was turning so fast there in the same direction that the shadow was moving that the eclipse lasted almost seven minutes at this equatorial location, about as long as an eclipse ever gets. Then the shadow passed over more southerly parts of Mexico, over Central America, and over parts of South America, before moving off the earth near sunset in Brazil.

"Which way will the sun be?" said Kenneth Eskaran, the mason, as he prepared bases of cinder blocks and cement for our telescopes. Our telescopes needed to be so securely mounted that they wouldn't shake even as they followed the sun across the sky. We arrived on site in Waikoloa, on the west coast of the Big Island of Hawaii, weeks in advance with our two tons

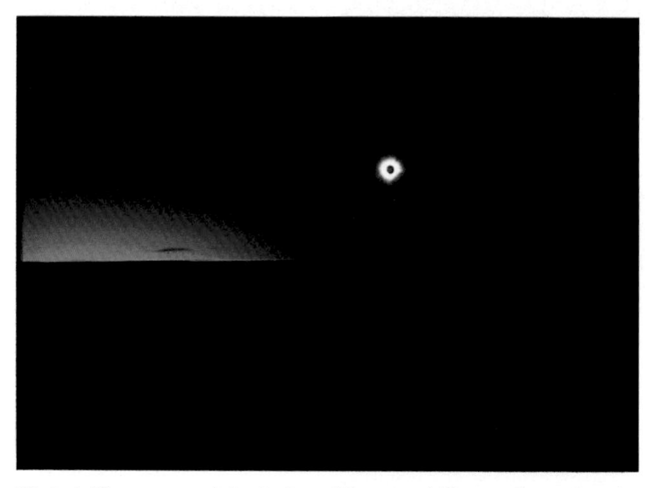

Plate 1. *The corona and the shadow of the moon falling on the earth at the 1990 eclipse, taken from an airplane over Finland with a 24-mm Nikkor lens and a Nikon EL2. (Deborah Pasachoff)*

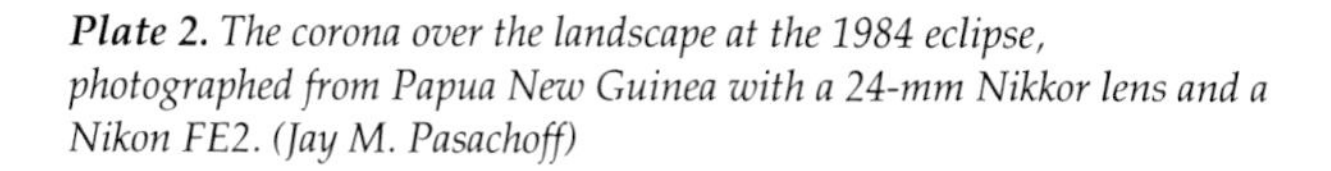

Plate 2. *The corona over the landscape at the 1984 eclipse, photographed from Papua New Guinea with a 24-mm Nikkor lens and a Nikon FE2. (Jay M. Pasachoff)*

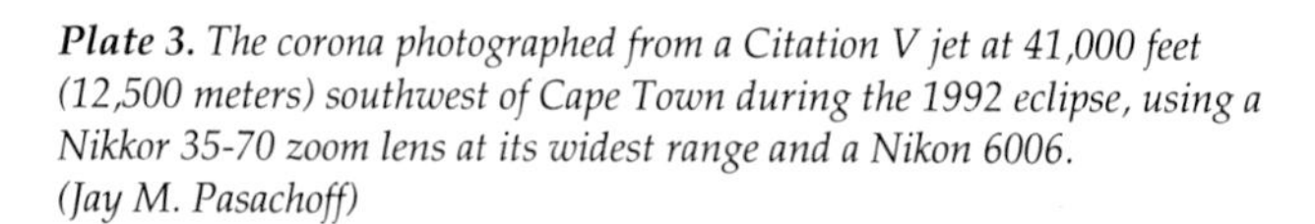

Plate 3. *The corona photographed from a Citation V jet at 41,000 feet (12,500 meters) southwest of Cape Town during the 1992 eclipse, using a Nikkor 35-70 zoom lens at its widest range and a Nikon 6006. (Jay M. Pasachoff)*

Plate 4. *The diamond ring and corona photographed from an airplane over the ocean near Hawaii during the 1981 eclipse with a 135-mm Nikkor telephoto lens on a Nikon F2as. The apparent 'feather' at the bottom of the sun is a scattering effect in the plastic airplane windows. (Jay M. Pasachoff)*

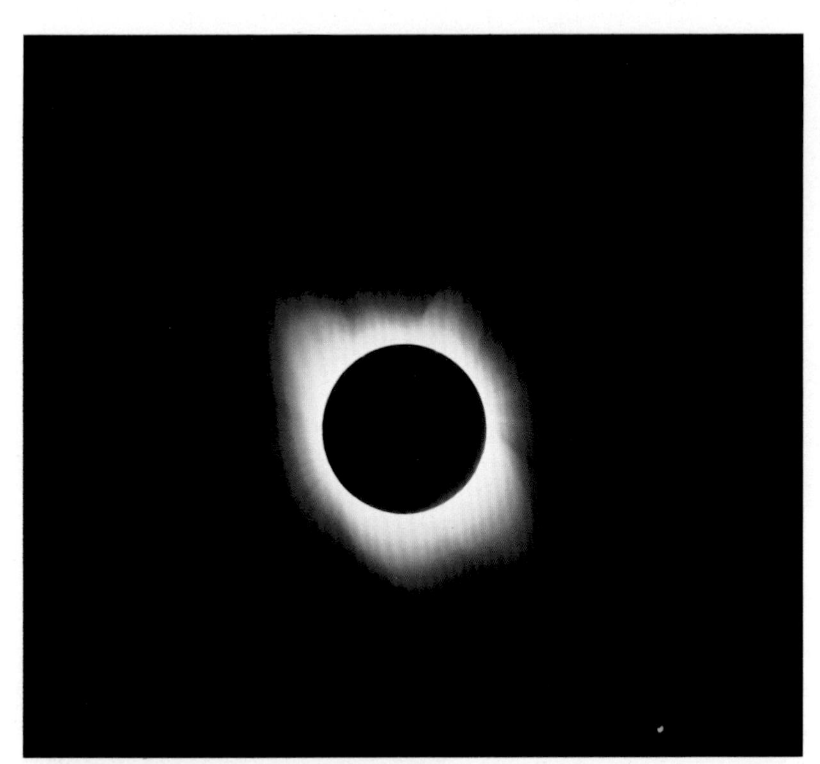

Plate 5. *The corona at the 1984 eclipse in Papua New Guinea, photographed on Kodachrome 64 film with a 500-mm Nikkor lens and a Nikon F2as. Note the asymmetric shape typical of solar minimum. (Jay M. Pasachoff)*

Plate 6. *The corona at the 1980 eclipse in India, photographed on Ektachrome 100 film with a 400-mm lens on a Mamaya large-scale camera. (Jay M. Pasachoff)*

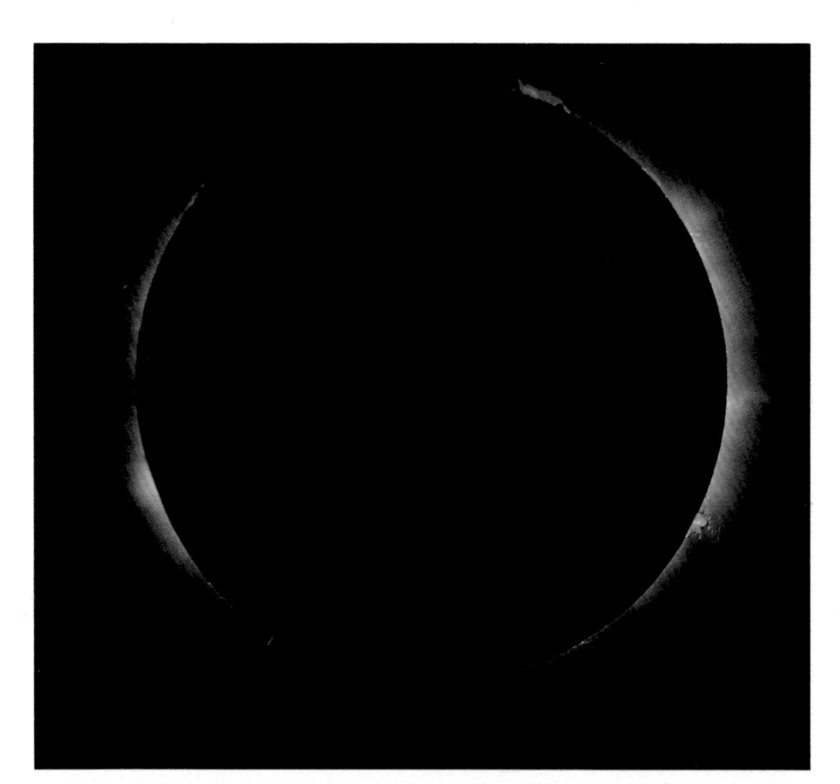

Plate 7. *Prominences at the edge of the sun and the innermost corona at the total solar eclipse of 1979, photographed through an 8″ (20-cm) Celestron telescope with an attached Nikon F2as camera body. (Jay M. Pasachoff and Martin Weinhous)*

Plate 8. *The diamond ring effect, photographed through an 8″ (20-cm) Celestron telescope with an attached Nikon FTn body. (Jay M. Pasachoff)*

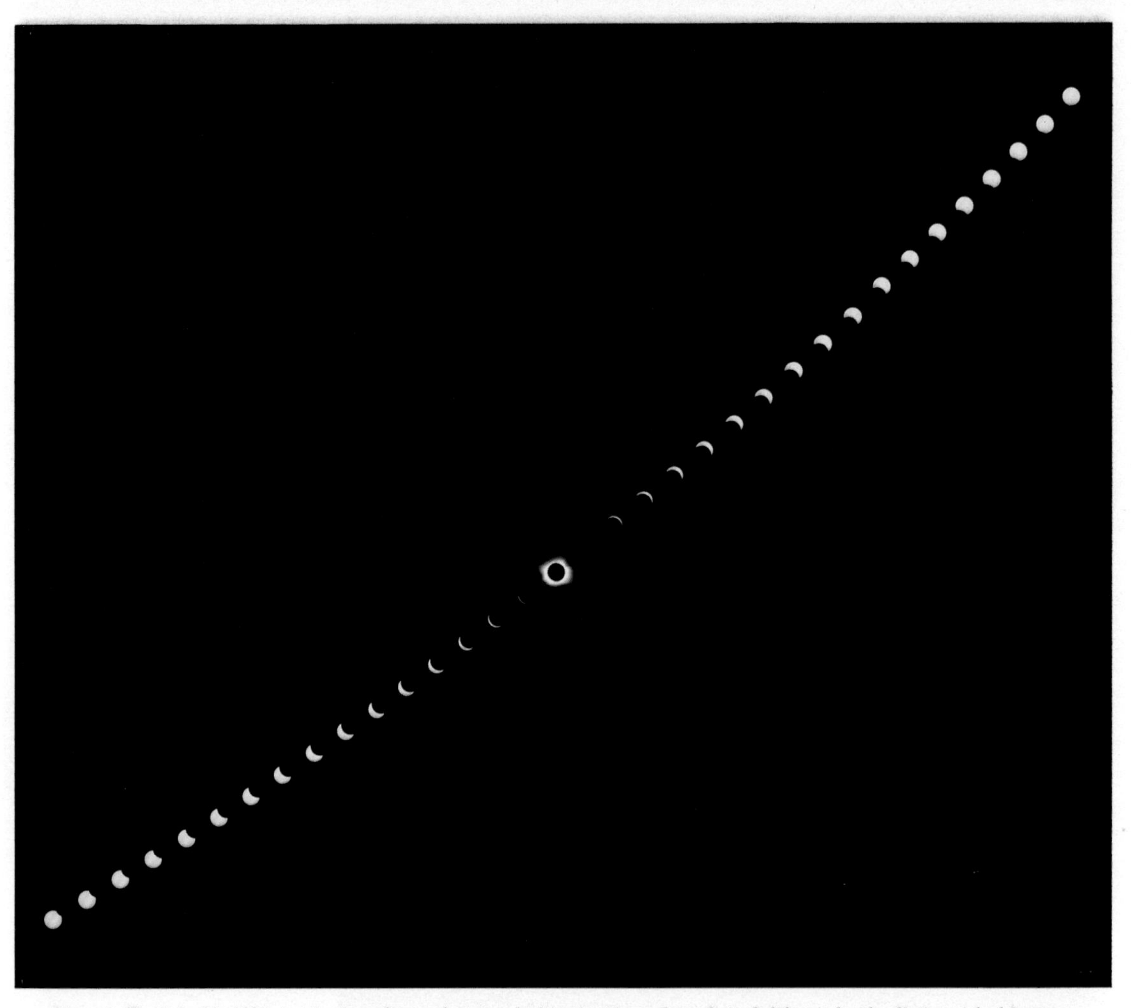

Plate 9. *The July 11, 1991, total solar eclipse, photographed in time lapse from Baja California by the distinguished Japanese amateur astronomer Akira Fujii. He used a Mamiya Press camera with a 100-mm lens; the interval between exposures was 5 minutes.*

Plate 10. *The annular eclipse of 1992, photographed from an. airplane off the coast of southern California at sunset through clouds with a 135-mm Nikkor telephoto and a Nikon 6006 camera. (Jay M. Pasachoff)*

Plate 11. *The annular eclipse of 1984, photographed from Mississippi without a filter using a 500-mm Nikkor lens on a Nikon F2as. (Jay M. Pasachoff)*
See also Figure 2.6.

Plate 12. *A time sequence showing the annular eclipse of 1973, photographed from Costa Rica. (Dennis di Cicco)*

***Plate 13.** A time-lapse sequence showing the moon rising and coming out of the 1982 lunar eclipse. Because of the different penetration of the moon into the umbra at different times, exposures vary between 1/2 second at f/5.6 for the thin crescent to 1/125 second at f/11 for the nearly full moon. (Akira Fujii)*

Plate 14. *The ruddy glow of the umbra contrasts with the bright white of the penumbra in this photograph of the lunar eclipse of December 9, 1992. The exposure was 30 seconds on Ektachrome 100 with a Celestron 8″ (20-cm) f/1.5 Schmidt camera. (Dennis di Cicco)*

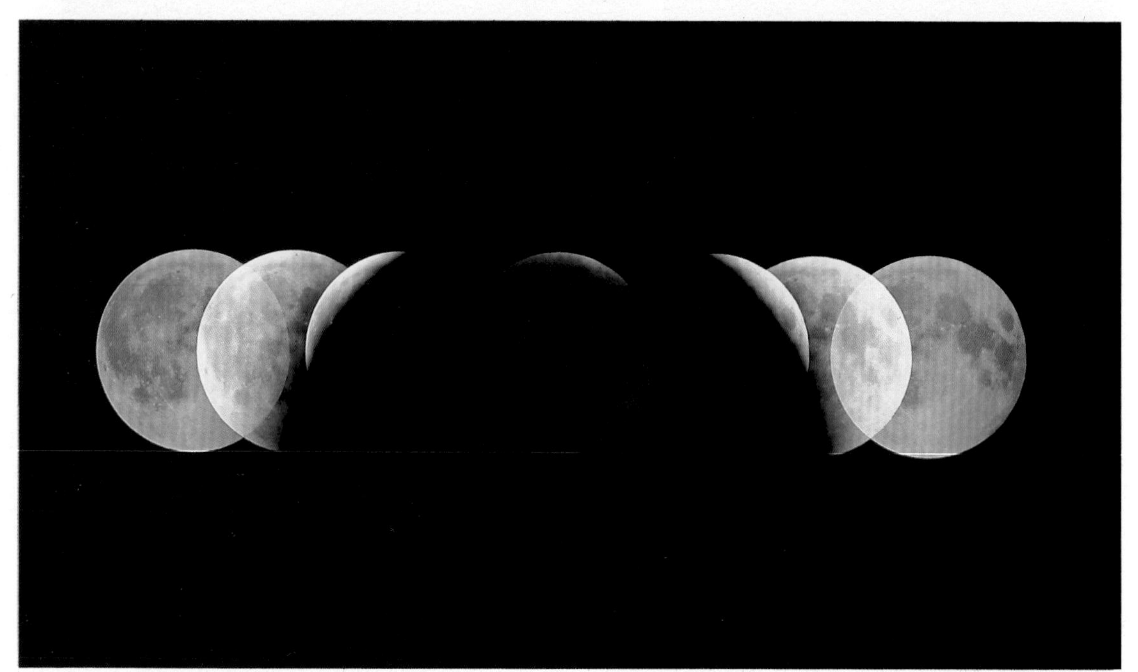

Plate 15. *The camera was tracked at the rate the stars move across the sky in this multiple exposure that reveals the shape of the earth's shadow in the sky. (Akira Fujii)*

of equipment. We were preparing to take the sun's temperature, part of finding out how the sun shines. Kevin Reardon, a Williams College senior, and I were to do so with the help of a CCD, a new type of electronic detector that is over 50 times more sensitive to light than film. (Reardon is now working at the Institute for Astronomy of the University of Hawaii.) As a result, what would have taken us 200 minutes of eclipse to observe 10 years ago, we can now observe in four minutes, which is a good thing, because four minutes is all we were to have.

"Why the solar corona is so hot is one of the most important unsolved problems in astronomy," said Peter Ulmschneider of the University of Heidelberg, Germany. Our second experiment was to search for a solution to this problem. Ulmschneider had called an international meeting on the single topic of the heating of the outer solar atmosphere, and over 100 astronomers showed up. "I don't have to worry about the weather, though," said Ulmschneider. "If it is cloudy, or if my computer doesn't work, I just run the program the next day."

We had run an earlier version of our second experiment at the eclipse in Java in 1983, but electronics had advanced a lot since then. "What you had to build from scratch then we can now do with off-the-shelf pieces and a Macintosh II," said Dr. Bryce Babcock, staff physicist at Williams College and my main electronics consultant. "Within a few weeks, Brad and I had arranged to take data every 1/1,000 of a second." Brad Behr was a senior at Williams College who came to Hawaii with us to work on the experiment, and at the time of writing, is working at Cambridge Research and Instrumentation in Cambridge, Mass. "And we recorded the data at four different amplifications to make sure we recorded them as well as possible."

On "the mountain," Mauna Kea, other scientists were also studying how the corona got to be millions of degrees. It seems strange to astronomers, just as it seems strange to everybody, that this million-degree gas should be on top of the solar surface, which is "only" 6 thousand degrees Celsius (11 thousand degrees Fahrenheit). Barry LaBonte of the University of Hawaii, using the 2.2-m (88-inch) telescope on Mauna Kea, took electronic images 16 times during the eclipse. "We hope to see little bright points go off, like firecrackers," LaBonte explained. "These microflares could heat the corona, if they exist."

Since the most important advance in astronomical instrumentation of the last few years is arrays able to make images in the infrared, six different groups planned to study the corona with them. Three of the groups were on Mauna Kea. "We were looking for a dust ring around the sun," explained Don Hall. "Jupiter and Saturn have rings, and perhaps the sun has one too. Dust glows in the infrared, so viewing in the infrared is the best way to study it."

As eclipse day approached, we each prepared and we each worried in his or her own way. A 90% chance of clear weather is good odds, but is no sure thing. Still, we had been waiting for so long, eight years since the last "major eclipse," an eclipse with favorable weather forecasts. And that eclipse had to be viewed through clouds. Would we do better this time?

As I awoke at 3 a.m., and 4 a.m., and 5 a.m. each morning to scan the skies, I was first worried and then relieved. There was more cloudiness in the Hawaiian morning sky than I had hoped. But the cloudiness was thin, and the sun almost always rose above it by eclipse time.

None of us were prepared for a major, new worry. We had all read about the devastating eruptions of Mt. Pinatubo in the Philippines only a month earlier. But nobody thought right away that the dust would girdle the earth. "We wish it had waited another week to arrive," said Hall. By a few days before the eclipse, the entire band of the earth between latitudes 15 and 25 degrees was covered with dust that made the sky look hazy. The dust was so high, in the stratosphere, that even Mauna Kea wasn't high enough to escape it. We could only hope that it didn't get worse. Looking through the dust in the earth's stratosphere would certainly make it more difficult to know if there was dust out in space.

As the eclipse approached, the mornings were clear enough. A couple of days before totality, scientists on Mauna Kea were allowed to install their instruments on the big telescopes. The nighttime astronomers didn't want to lose observing time to mere daylight astronomers. The day before the eclipse, as we all rehearsed our techniques, it was clear in Hawaii. But Baja California was under clouds.

Finally, July 11 arrived. All was ready. Our telescopes had never tracked the path of the sun so accurately. All our electronics was also working perfectly. Our programs were primed to run in the four minutes and seven seconds of totality we would have. We could look up and see the domes of the telescopes on Mauna Kea, 35 miles (55 kilometers) to our east. We knew that our colleagues were also setting up their equipment in those domes. The sunset the night before had been fantastically red, something we would have ordinarily thought was beautiful. Now we could think only of the Mt. Pinatubo dust that had travelled around the world to make the extraordinary color.

At dawn, clouds covered the eastern part of the sky. Would the clouds disperse, perhaps when the eclipse cooled the atmosphere? Would the clouds retreat below 20° above the horizon, allowing us to see the eclipse above them? We were worried but hopeful as we started our procedures. We made our calibration exposures and started our tests.

With a half hour to go, "It's going to clear," I shouted to the assembled multitude of colleagues and students. The clouds were retreating below 20°. Would they stay there? Domes on Mauna Kea became visible in the distance.

On Mauna Kea, the scientists had their own worries. Fog was rising and nipping at the bases of the domes. Cirrus cloud covered the sky, but not so

Fig. 8.1. *The partially eclipsed sun, visible through clouds on July 11, 1991, from JMP's site in Hawaii. The photograph was taken with a Nikon F4 and a Nikkor 500-mm f/8 telephoto mirror lens; no filter was used because the clouds and haze provided the right amount of filtering. Still, we did not look through the viewfinder; the camera had been mounted so that it was tracking the sun.*

thickly as to obscure the sun. And everybody knew that a high level of volcanic dust was waiting to scatter light around, confusing everything.

As the minutes went by, our euphoria waned. The clouds that had been retreating regrouped, and began to climb again across the sky. A faint hole in the clouds appeared far to our south and behind us, probably, we thought, over the ocean. In any case, our carefully mounted equipment wasn't portable.

The last moments before totality are exciting, whether or not you can see the corona. The sky, which has been slowly growing darker, darkens dramatically and quickly. We knew we were in the moon's shadow. "Diamond ring," I called out, hoping that a hole in the clouds would open and we could see what we knew was happening. But it was not to be. Still, we could see the distant horizon appear ghostly pale, since we were seeing so far away that points there were out of the moon's shadow. "Awesome," said Eric Kutner, my Princeton-bound nephew. "Awesome, but disappointing." We counted the seconds. Deborah Pasachoff, a student at Deerfield Academy (Deerfield, Massachusetts), photographed the sky with a wide-angle lens, documenting the change in sky color. Four minutes is a tremendously long time if you are waiting and counting. Then, with a rush, the sky brightened. The total eclipse was over for us. We had proved, experimentally, that 90% isn't 100%.

On the mountain, the eclipse arrived a few seconds after it reached us. The fog had retreated. The corona was visible through a veil of clouds. Exposures might be a little longer, and light might be scattered around from one part of the corona to another, but most of the experiments would be OK.

"Could you see the eclipse at all?" I asked. "There were three of us astronomers in the control room," said Jean-Claude Vial of the Institut d'Astrophysique Spatiale in Verrières-le-Buisson, which has just moved to Orsay, France. He was one of the scientists using the giant Canada–France–Hawaii telescope, with its 3.6-m-diameter mirror. "Serge Koutchmy, of course, and Guy Monnet, the Director of the CFH, were also in the room." But Vial, Koutchmy, and colleagues had five different cameras on the telescope. Some used large-size film while others were electronic. "I was there, operating the Hasselblad," a camera attached to the telescope far overhead. "Before and after the actual times, a technician from Sac Peak (the solar observatory), Roy Coulter, checked the fast camera, and during totality he took pictures. And Serge said 'go' and 'stop.'" "That's an important job," I said. "At my site, I also had responsibility for deciding when the eclipse began." The French team has lots of images of the solar corona to study.

To my question, "are you sorry you didn't see the eclipse?" Vial laughs. "I was very stressed. My first basic impression is that time was running very fast. I was on the Finland team in the plane last year, which was even worse because duration was only 20 seconds. It was not panic, but something close. Four minutes goes too fast. I am sorry not to have seen the eclipse in real time, but . . . ." Vial tries to reassure me. "It was stress, but it was fun. After first contact [when the moon first touches the sun], I was on the catwalk and had a beautiful view of the landscape and the other astronomers, and it was very impressive to see all the other astronomers around the dome."

Nearby, in the 2.2-m telescope, LaBonte saw 16 beautiful partial images of the corona flash by on his television monitor. The next day, though, he was less happy. Somehow in the computer software that transferred data, the first image was recorded over and over again and the later strips were not recorded at all. Instead of writing sixteen strips, each 64 × 1,024 pixels in size, to make a final image 1,024 pixels square, the first strip was written sixteen times. The rest of the data, viewed so fleetingly on the computer monitor, were lost. Still, LaBonte and colleagues could study the data they had, and they have done so very carefully. No sign of microflares has turned up.

Two new kinds of telescopes worked perfectly on Mauna Kea. They study radiation on the boundary between long infrared and short radio waves. The waves of 1.3-mm wavelength measured by the James Clerk Maxwell telescope told Drake Deming and Charlie Lindsay about a layer of the solar atmosphere just below the corona, a layer otherwise impossible to study well. The waves of 0.85 millimeter told Hal Zirin of Caltech about a nearby layer. Together, these two experiments improve our three-dimensional probing of the sun.

The infrared arrays all functioned well, "but the data will be difficult to interpret because of the Mt. Pinatubo dust," explained Hall just after the eclipse. Two weeks later, he was happier. "We can definitely rule out a dust ring," he said. By a year later, several groups had analyzed their data, and these final results agree: no dust ring. Why didn't the dust ring reported a dozen years earlier show up? Perhaps it had been temporary, possibly caused by a comet that had passed recently.

Two of the telescopes on Mauna Kea couldn't study the eclipse at all. The then largest telescope, the United Kingdom Infrared Telescope, wasn't designed to point low enough in the sky. And the Keck Telescope, now the largest in the world, still had a month to go before enough glass mirrors were inserted to make it so. It will have to provide astronomical glories on objects fainter than the sun.

Attendance on Mauna Kea was kept low during the eclipse, though celebrities like the Governor of Hawaii, John Waihee, were allowed up. *National Geographic* photographer Roger Ressmeyer along with editors of *National Geographic* were there. Also, the distinguished solar astronomer Eijiro Hiei from the National Observatory of Japan, Tokyo, was present. Hiei, my predecessor as Chair of the Working Group on Eclipses of the International Astronomical Union, supervised a team from the national Japanese television network who made a High Definition Television recording of totality. At the IAU General Assembly two weeks later in Buenos Aires, Hiei explained, as he showed the result, "We had three high-definition cameras. One TV camera took the whole corona, one took part of the corona, a half turn from the east limb to the west limb, and one took the corona with the surrounding horizon. The remarkable, bright prominence that stood out far above the sun was well recorded." "What did you do during the eclipse," I asked? "I looked at the fine structure of the corona using a Questar. I looked especially at the prominence and the surrounding corona. I looked especially to see the fine structure of the coronal streamers. The corona is composed of many string-like structures. When I saw it this time, the strings were very fine. Sometimes they are fuzzy. At this eclipse, I could not see any dark region around the prominence." Such dark regions sometimes separate a prominence from the corona. Did he enjoy the eclipse? "Yes, of course. Especially, this eclipse was very beautiful. There were active prominences on the east side and on the west side. The corona was quite extended. Usually, I see an eclipse for only 10 seconds or so because I am busy with an experiment, but this time I really enjoyed it from the beginning to the end." An IMAX camera was also present, photographing onto giant 70-mm film.

One of the wonders of the eclipse was yet to come as totality ended on Mauna Kea. A look off the mountain to the east showed a dark hole – the moon's shadow below as it moved off toward Mexico. And weather satellites from both Japan and the United States recorded the progress of the moon's elliptical shadow across the earth.

By a half hour after totality ended in Hawaii, we could see the partially eclipsed sun through the clouds. As it played peek-a-boo, Wellesley student Michelle Freddolino and Harvard student Eloise Pasachoff photographed the partially eclipsed sun with the cameras they had hoped to use on totality. Soon thereafter, the sun had risen above the clouds, and the clouds were dispersing. By 8 a.m. local time, the sky was clear. The day would go down as a clear one in the official weather records of Hawaii.

While the eclipse was in progress in Hawaii, Leon Golub of the Harvard–Smithsonian Center for Astrophysics launched a rocket from White Sands, New Mexico. Rather, NASA launched it for him. Golub has built a camera that can photograph the x-rays from the solar corona right in front of the sun, without need for the eclipse. We can connect the structures Golub sees on the sun's disk with the structures seen in the corona to the sides of the disk at the eclipse. All Golub's x-ray images are taken at the same x-ray energy, which shows gas of the same temperature. As Carole Jordan of the University of Oxford explains, "active regions are composed of loops of different minimum temperatures. The rocket observations show us only one temperature. It would clearly be desirable to have such high-quality data at a wide range of temperatures." We will have to await more rockets and more eclipses.

Two hours later, the eclipse reached Baja California. Twenty-five thousand tourists and a hundred or so professional astronomers were gathered there near La Paz. "I had never seen an eclipse before," said Julietta Fierro, "so it was outstanding. I was very pleased because organizing this campsite in Mexico, I was extremely frightened that things wouldn't work out. There was enough electricity, and most people were happy."

"What did you think of the eclipse?" I asked. "I felt amazed by our universe. I felt overwhelmed. I felt surprised by how accurate the predictions were – everything people said would happen, happened. It was so beautiful. I just lay there on the floor with my two sons surrounded by this beauty of nature. It was amazing.

"I think the scientists were successful. There were 17 countries and 16 experiments. They were successful. That made me very happy because we worked on this campsite for several years. Since the day before was cloudy when we held the press conference, I could see the faces that looked so disappointed, especially the people from Japan. I saw their sad, sad faces. I had tried to see an annular eclipse in Mexico before but it was cloudy, and I feared the same thing would happen. At 11 or 12 at night, it started clearing, and the morning of the eclipse I was awakened by the birds. They were singing so loud I rushed out of my tent and the sky was so blue. It was a warm feeling. It was happiness."

"How did you like the corona?" "Oh, did I! It was so beautiful. Experts said it was big, but since I didn't know what to expect, it baffled my experiences. I couldn't believe the prominences. It was like a dream come true."

In Baja California, it was cloudy the day before and cloudy the day after, but eclipse day was fine. Eclipse veteran Steve Edberg, of the Jet Propulsion Laboratory, said "I hate to tell you this, Jay, but it was the best eclipse I have been to. The corona was spectacular." In Waikoloa, Hawaii, on the other hand, 18 of the twenty days before and during the eclipse were sunny, so the weather was right on the predictions of 90% chance of clear. But as luck would have it, the 10% included eclipse day. On the Hawaiian island of Maui, for which a sliver of coast was in totality, Judith Kissell told me, "Do you know Kaupo, past Hana? They expected 40,000 people in Kaupo on that one-lane road, but with the weather, not many people came, so it was easy to get to. It was rainy. It poured! It just doesn't do that in the summer! Even Haleakala was socked in."

As for Mexico, farther down the coast than in Baja, many people were in cars maneuvering for holes in the clouds near Mazatlán. Some were successful and others not. In Mexico City, millions of people were able to see totality. "From my house, up to 10 minutes before totality, it was cloudy," said Dr. Alexandro Ruelas of the National University of Mexico. "It was quite thick; you couldn't see the sun. Then it cleared up and remained clear until one minute into totality. It was cloudy for the rest of totality and I didn't get to see the sun for another half hour." As for the eclipse, "It was magnificent. I enjoyed every second of it. The thing that was most impressive was the diamond ring." He went on, "The weather pattern was strange. Fifty miles south in Cuernavaca, it was completely clear."

The ancient Maya, over a thousand years ago, had known that eclipses could occur every 177 days. Art historian Sam Edgerton, who teaches about the Maya, was on a beach in Costa Rica. "We could stand on the sand and see totality," he reported. Edgerton is at the Clark Art Institute in Williamstown, Massachusetts, and at Williams College.

Koutchmy, in addition to his work on Mauna Kea, had arranged for six identical telescopes to be spaced along the eclipse path. Two were in Hawaii, two in Mexico, and two in Brazil. The cameras should show any changes in the corona, even small ones, over a span of a few hours. But no changes were seen. It is interesting to note that the camera that was under his direct control, on Mauna Kea, gave the best results. The other camera in Hawaii was clouded out. As for the pair of cameras in Mexico, the film from one was lost in an accident during the developing process, and the other camera was successful, though the image and the radial filter were off center, making it more difficult to interpret than one would like. An image from Brazil is OK.

All across the Americas, people could see a partial eclipse. The moon was off to the side of the sun so didn't cover the sun entirely. Most people don't realize that the everyday sun is one million times brighter than the corona. So even when 99% of the sun is covered, we have 10,000 times more light from the sliver of sun remaining than comes from the corona. Thus 99% coverage is only 0.0001 of the way from full sun toward totality. If we called it a 1/100% coverage instead of 99% coverage, people would more accurately understand what is occurring.

This 1991 eclipse was also described as "the eclipse of the century." It was perhaps the fourth or fifth I have heard described that way, so perhaps I am a bit skeptical. There is a total solar eclipse, after all, about every year, though one usually has to travel to it. Often, the distinction between total eclipse and partial eclipse is not carefully made. The people of Honolulu, who missed totality this time, have been told that there won't be another "eclipse" in Hawaii for more than a hundred years. But, indeed, they have already had another partial eclipse, on January 4, 1992, when 70% of the sun was covered from Honolulu. I hope they weren't too confused.

I have since seen the annular eclipse of January 4, 1992, at sunset over the Pacific Ocean from the coast of California off San Diego. And I have seen the total eclipse of June 30, 1992, from an airplane off the coast of Africa. This book details the further total and annular eclipses of this decade. I'm looking forward to them all.

APPENDIX A

SOURCES OF FURTHER INFORMATION

Parts of this book are taken from *Astrophotography for the Amateur*, by Michael Covington (Cambridge University Press, 1991); further references are given there. The following are some information sources especially relevant to eclipses.

Magazines

Sky and Telescope, 49 Bay State Road, Cambridge, Massachusetts 02138, USA, (800) 253-0245 or (617) 864-7360. (Extensive articles about each major eclipse before and after it.)

Astronomy, 21027 Crossroads Circle, Waukesha, Wisconsin 53187, USA, (800) 533-6644.

Journal of the British Astronomical Association, Burlington House, Piccadilly, London W1V 9AG, England.

Astronomy Now, 193 Uxbridge Road, London W12 9RA, England, (081) 743 8888.

Eclipse predictions

Canon of Solar Eclipses: 1986–2035, by Fred Espenak, NASA Reference Publication 1178 (Revised, 1987). Available from Willmann-Bell. The best set of maps and path computations.

Astronomical Almanac and *Astronomical Phenomena*. Published annually by US Government Printing Office, Washington, DC 20402, and HM Stationery Office, PO Box 276, London SW8 5DT. Of the two, *Phenomena* is a shorter book and is available farther in advance. Both contain large-area maps and predictions (in tables like Tables 4.1 and 6.1) for eclipses during the relevant year. Libraries often shelve these books with government publications rather than astronomy.

Circular No. 170, US Naval Observatory, Washington, DC 20390. Large-area maps and tables for all solar eclipses through the year 2000. The US Naval Observatory has stopped issuing detailed circulars about individual eclipses.

Astronomical Tables of the Sun, Moon, and Planets, by Jean Meeus. Richmond (Virginia): Willmann-Bell (PO Box 35025, Richmond, VA 23235, (804) 320-7016), 1983. Brief information on all eclipses through the year 2050.

Astronomer Fred Espenak (NASA/Goddard Space Flight Center) and meteorologist Jay Anderson (Prairie Weather Center, Winnipeg) are producing and distributing a series of eclipse bulletins as part of NASA's Reference Publication series. These bulletins are the successors to the discontinued Circulars of the US Naval Observatory. The first such Reference Publication deals with the 1994 annular eclipse and the second deals with the 1994 total eclipse. For single copies, please write:
Jay Anderson, Prairie Weather Center, 900-266 Graham Avenue, Winnipeg MB, Canada R3C 3V4.

Observing techniques

Field Guide to the Stars and Planets, 3rd ed., by Jay M. Pasachoff and Donald H. Menzel, Peterson Field Guide Series, Boston: Houghton Mifflin and London: Victor Gollancz, 1992.

Peterson First Guide to Astronomy and *Peterson First Guide to the Solar System*, by Jay M. Pasachoff, Boston: Houghton Mifflin, 1989 and 1990.

The Guide to Amateur Astronomy, by Jack Newton and Phillip Teece, Cambridge University Press, 1988.

Observing the Sun, by Peter O. Taylor, Cambridge University Press, 1991.

National Geographic articles about total eclipses

Old *National Geographic* articles about eclipses and eclipse expeditions are fascinating. Those planning to photograph eclipses can view a wide variety of photographs, including many multiple-shot time-lapse views.

May 1992 issue: "The great eclipse" and "The darkness that enlightens," by Jay M. Pasachoff and Roger Ressmeyer, vol. 181, no. 5, pp. 30–51; about the July 11, 1991, eclipse in Hawaii and Mexico.

August 1970 issue: "Solar eclipse: nature's super spectacular," by Donald H. Menzel and Jay M. Pasachoff, vol. 138, no. 3, pp. 222–3; about the March 7, 1970, eclipse in Mexico.

November 1963 issue: "The solar eclipse from a jet," by Wolfgang B. Klemperer, pp. 785–96; July 20, 1963, eclipse, with the path from Hokkaido, Japan, to beyond Bar Harbor, Maine.

February 1953 issue: "South in the Sudan," by Harry Hoogstraal, an article about Sudan including the Khartoum eclipse site, vol. 103, pp. 251–2, and a further picture of an astronomer with his equipment, p. 706, about the February 25, 1952, eclipse.

March 1949 issue: "Operation Eclipse: 1948," by William A. Kinney, vol. 95, pp. 325–72, with 35 color illustrations; about the May 8–9, 1948, eclipse.

September 1947 issue: "Eclipse hunting in Brazil's Ranchland," by F. Barrows Colton, vol. 92, pp. 285–324 + 16 pp. of color plates; about the May 20, 1947, eclipse.

May 1947 issue: "Your society observes eclipse in Brazil," vol. 91, p. 661; an advance announcement of the May 20, 1947, eclipse.

September 1937 issue: "Eclipse adventures on a desert isle," by J. F. Hellweg, vol. 72, pp. 377–94, and "Nature's most dramatic spectacle," by S. A. Mitchell, vol. 72, pp. 361–76; about the Pacific Ocean eclipse of June 8, 1937, that passed Canton Island.

February 1937 issue: "Observing an eclipse in Asiatic Russia," by Irvine C. Gardner, with "the first natural-color photograph of a total eclipse ever published," vol. 71, pp. 178–97; about the June 19, 1936, eclipse in Ak Bulak, Kazakhstan.

November 1932 issue: "Photographing the eclipse of 1932 from the air," by Albert W. Stevens, vol. 62, pp. 581–96, and "Observing a total eclipse of the sun," by Paul A. McNally, vol. 62, pp. 597–605; eclipse of August 31, 1932, over New England.

Scientific background

Astronomy: From the Earth to the Universe, 4th ed., 1993 version, by Jay M. Pasachoff. Philadelphia: Saunders College Publishing, 1993. A clear explanation of all parts of astronomy, including the sun and eclipses and what they tell us. Beautifully illustrated, completely in color. Order from your college bookstore or from Astronomy Book, 1305 Main Street, Williamstown, MA 01267. A fifth edition is scheduled for 1995.

Filters

Aluminized Mylar
Roger Tuthill, Inc., Box 1086, Mountainside, NJ 07092, (800) 223-1063 or (908) 232-1786.

Aluminized glass
Thousand Oaks Optical, Box 248098, Farmington, MI 48332-8098, (313) 353-6825.

Solar filters are also available from most telescope manufacturers.

Video game

"Where in Space is Carmen Sandiego?" (Broderbund Software).
Each box includes the *Peterson First Guide to Astronomy* by Jay M. Pasachoff.

Appendix B

Exposure tables

These exposure tables are selected from those in *Astrophotography for the Amateur,* which contains full explanations of how they were calculated, as well as tables for other celestial objects.

These exposures are only approximate. Because the transparency of the air is highly variable, all exposures should be bracketed at least 1 stop and preferably 2 stops either side of the value in the table. Some experts prefer different values. For example, Dennis di Cicco prefers 1/125 second to the 1/15 second exposure time given here for prominences for *f*/16 and ISO (ASA) 64.

Here "≪" denotes an exposure of less than 1/1,000 second (too short for most shutters), and "≫" denotes an exposure longer than five minutes (and hence impossible to predict accurately because of variations in reciprocity failure, a film effect in which the sensitivity of the film to incident light varies with the length of the exposure).

B is a constant that indicates the brightness of the object being photographed. The uneclipsed full moon has a *B* value of 200. The uses of *B* values are described on p. 51.

MOON – thin crescent
$B = 10$

f/	ISO (ASA) 32	ISO (ASA) 64	ISO (ASA) 100	ISO (ASA) 200	ISO (ASA) 400
2	1/60	1/125	1/250	1/500	1/1,000
2.8	1/30	1/60	1/125	1/250	1/500
4	1/15	1/30	1/60	1/125	1/250
5.6	1/8	1/15	1/30	1/60	1/125
8	1/4	1/8	1/15	1/30	1/60
11	1/2	1/4	1/8	1/15	1/30
16	1	1/2	1/4	1/8	1/15
22	3	1	1/2	1/4	1/8
32	6	3	2	1/2	1/4
45	15	6	4	2	1/2
64	40	15	9	4	2
100	130	50	30	11	5
130	265	100	55	20	9
160	≫	180	100	40	15
200	≫	≫	180	70	30
250	≫	≫	≫	130	50
300	≫	≫	≫	210	80

MOON – wide crescent
Also: dimly lit features on terminator at any time
$B = 20$

f/	ISO (ASA) 32	ISO (ASA) 64	ISO (ASA) 100	ISO (ASA) 200	ISO (ASA) 400
2	1/125	1/250	1/500	1/1,000	1/2,000
2.8	1/60	1/125	1/250	1/500	1/1,000
4	1/30	1/60	1/125	1/250	1/500
5.6	1/15	1/30	1/60	1/125	1/250
8	1/8	1/15	1/30	1/60	1/125
11	1/4	1/8	1/15	1/30	1/60
16	1/2	1/4	1/8	1/15	1/30
22	1	1/2	1/4	1/8	1/15
32	3	1	1/2	1/4	1/8
45	6	3	2	1/2	1/4
64	15	6	4	2	1/2
100	50	20	11	5	2
130	100	40	20	9	4
160	180	70	40	15	6
200	≫	130	70	30	11
250	≫	235	130	50	20
300	≫	≫	210	80	30

MOON – quarter phase

Also: brightly lit features on terminator at any time

$B = 40$

f/	ISO (ASA) 32	ISO (ASA) 64	ISO (ASA) 100	ISO (ASA) 200	ISO (ASA) 400
2	1/250	1/500	1/1,000	1/2,000	≪
2.8	1/125	1/250	1/500	1/1,000	1/2,000
4	1/60	1/125	1/250	1/500	1/1,000
5.6	1/30	1/60	1/125	1/250	1/500
8	1/15	1/30	1/60	1/125	1/250
11	1/8	1/15	1/30	1/60	1/125
16	1/4	1/8	1/15	1/30	1/60
22	1/2	1/4	1/8	1/15	1/30
32	1	1/2	1/4	1/8	1/15
45	3	1	1/2	1/4	1/8
64	6	3	2	1/2	1/4
100	20	8	5	2	1/2
130	40	16	9	4	2
160	70	30	15	6	3
200	130	50	30	11	5
250	235	90	50	20	8
300	≫	150	80	30	13

MOON – gibbous

$B = 80$

f/	ISO (ASA) 32	ISO (ASA) 64	ISO (ASA) 100	ISO (ASA) 200	ISO (ASA) 400
2	1/500	1/1,000	1/2,000	≪	≪
2.8	1/250	1/500	1/1,000	1/2,000	≪
4	1/125	1/250	1/500	1/1,000	1/2,000
5.6	1/60	1/125	1/250	1/500	1/1,000
8	1/30	1/60	1/125	1/250	1/500
11	1/15	1/30	1/60	1/125	1/250
16	1/8	1/15	1/30	1/60	1/125
22	1/4	1/8	1/15	1/30	1/60
32	1/2	1/4	1/8	1/15	1/30
45	1	1/2	1/4	1/8	1/15
64	3	1	1/2	1/4	1/8
100	8	4	2	1/2	1/4
130	16	7	4	2	1/2
160	30	11	6	3	1
200	50	20	11	5	2
250	90	35	20	8	4
300	150	60	30	13	6

MOON – full
$B = 200$

f/	ISO (ASA) 32	ISO (ASA) 64	ISO (ASA) 100	ISO (ASA) 200	ISO (ASA) 400
2	1/2,000	≪	≪	≪	≪
2.8	1/1,000	1/2,000	1/2,000	≪	≪
4	1/500	1/1,000	1/1,000	1/2,000	≪
5.6	1/250	1/500	1/500	1/1,000	1/2,000
8	1/125	1/250	1/250	1/500	1/1,000
11	1/60	1/125	1/125	1/250	1/500
16	1/30	1/60	1/60	1/125	1/250
22	1/15	1/30	1/30	1/60	1/125
32	1/8	1/15	1/15	1/30	1/60
45	1/4	1/8	1/8	1/15	1/30
64	1/2	1/4	1/4	1/8	1/15
100	3	1	1/2	1/4	1/8
130	5	2	1	1/2	1/4
160	9	4	2	1/2	1/4
200	15	6	4	2	1/2
250	25	11	6	3	1
300	45	17	10	4	2

MOON – partial eclipse
Exposing for penumbra (light portion) only
$B = 50$

f/	ISO (ASA) 32	ISO (ASA) 64	ISO (ASA) 100	ISO (ASA) 200	ISO (ASA) 400
2	1/500	1/1,000	1/1,000	1/2,000	≪
2.8	1/250	1/500	1/500	1/1,000	1/2,000
4	1/125	1/250	1/250	1/500	1/1,000
5.6	1/60	1/125	1/125	1/250	1/500
8	1/30	1/60	1/60	1/125	1/250
11	1/15	1/30	1/30	1/60	1/125
16	1/8	1/15	1/15	1/30	1/60
22	1/4	1/8	1/8	1/15	1/30
32	1/2	1/4	1/4	1/8	1/15
45	2	1/2	1/2	1/4	1/8
64	5	2	1	1/2	1/4
100	15	6	4	2	1/2
130	30	12	7	3	1
160	50	20	12	5	2
200	95	35	20	9	4
250	175	70	35	15	6
300	290	110	60	25	10

MOON – partial eclipse: umbra and penumbra together

Try a wide range of exposures – these are only suggestions
$B = 0.25$

f/	ISO (ASA) 32	ISO (ASA) 64	ISO (ASA) 100	ISO (ASA) 200	ISO (ASA) 400
2	1/2	1/4	1/8	1/15	1/30
2.8	1	1/2	1/4	1/8	1/15
4	4	2	1/2	1/4	1/8
5.6	8	4	2	1/2	1/4
8	20	9	5	2	1/2
11	50	19	11	5	2
16	135	50	30	12	5
22	≫	125	65	25	11
32	≫	≫	185	70	30
45	≫	≫	≫	185	70
64	≫	≫	≫	≫	185
100	≫	≫	≫	≫	≫
130	≫	≫	≫	≫	≫
160	≫	≫	≫	≫	≫
200	≫	≫	≫	≫	≫
250	≫	≫	≫	≫	≫
300	≫	≫	≫	≫	≫

MOON – relatively light total eclipse

Try a wide range of exposures – these are only suggestions
$B = 0.05$

f/	ISO (ASA) 32	ISO (ASA) 64	ISO (ASA) 100	ISO (ASA) 200	ISO (ASA) 400
2	5	2	1	1/2	1/4
2.8	11	5	3	1	1/2
4	30	11	6	3	1
5.6	70	25	15	6	3
8	180	70	40	15	6
11	≫	165	90	35	14
16	≫	≫	255	100	40
22	≫	≫	≫	235	90
32	≫	≫	≫	≫	255
45	≫	≫	≫	≫	≫
64	≫	≫	≫	≫	≫
100	≫	≫	≫	≫	≫
130	≫	≫	≫	≫	≫
160	≫	≫	≫	≫	≫
200	≫	≫	≫	≫	≫
250	≫	≫	≫	≫	≫
300	≫	≫	≫	≫	≫

MOON – relatively dark total eclipse

Try a wide range of exposures – these are only suggestions
$B = 0.005$

f/	ISO (ASA) 32	ISO (ASA) 64	ISO (ASA) 100	ISO (ASA) 200	ISO (ASA) 400
2	95	35	20	9	4
2.8	240	90	50	20	8
4	≫	245	135	50	20
5.6	≫	≫	≫	130	50
8	≫	≫	≫	≫	135
11	≫	≫	≫	≫	≫
16	≫	≫	≫	≫	≫
22	≫	≫	≫	≫	≫
32	≫	≫	≫	≫	≫
45	≫	≫	≫	≫	≫
64	≫	≫	≫	≫	≫
100	≫	≫	≫	≫	≫
130	≫	≫	≫	≫	≫
160	≫	≫	≫	≫	≫
200	≫	≫	≫	≫	≫
250	≫	≫	≫	≫	≫
300	≫	≫	≫	≫	≫

SUN – full disk or partial eclipse

Through full aperture Solar Skreen® filter
$B = 80$

f/	ISO (ASA) 32	ISO (ASA) 64	ISO (ASA) 100	ISO (ASA) 200	ISO (ASA) 400
2	1/500	1/1,000	1/2,000	≪	≪
2.8	1/250	1/500	1/1,000	1/2,000	≪
4	1/125	1/250	1/500	1/1,000	1/2,000
5.6	1/60	1/125	1/250	1/500	1/1,000
8	1/30	1/60	1/125	1/250	1/500
11	1/15	1/30	1/60	1/125	1/250
16	1/8	1/15	1/30	1/60	1/125
22	1/4	1/8	1/15	1/30	1/60
32	1/2	1/4	1/8	1/15	1/30
45	1	1/2	1/4	1/8	1/15
64	3	1	1/2	1/4	1/8
100	8	4	2	1/2	1/4
130	16	7	4	2	1/2
160	30	11	6	3	1
200	50	20	11	5	2
250	90	35	20	8	4
300	150	60	30	13	6

SUN – total eclipse: prominences
No filter
$B = 50$

f/	ISO (ASA) 32	ISO (ASA) 64	ISO (ASA) 100	ISO (ASA) 200	ISO (ASA) 400
2	1/500	1/1,000	1/1,000	1/2,000	≪
2.8	1/250	1/500	1/500	1/1,000	1/2,000
4	1/125	1/250	1/250	1/500	1/1,000
5.6	1/60	1/125	1/125	1/250	1/500
8	1/30	1/60	1/60	1/125	1/250
11	1/15	1/30	1/30	1/60	1/125
16	1/8	1/15	1/15	1/30	1/60
22	1/4	1/8	1/8	1/15	1/30
32	1/2	1/4	1/4	1/8	1/15
45	2	1/2	1/2	1/4	1/8
64	5	2	1	1/2	1/4
100	15	6	4	2	1/2
130	30	12	7	3	1
160	50	20	12	5	2
200	95	35	20	9	4
250	175	70	35	15	6
300	290	110	60	25	10

SUN – total eclipse: inner corona (3° field)
No filter
$B = 5$

f/	ISO (ASA) 32	ISO (ASA) 64	ISO (ASA) 100	ISO (ASA) 200	ISO (ASA) 400
2	1/30	1/60	1/125	1/250	1/500
2.8	1/15	1/30	1/60	1/125	1/250
4	1/8	1/15	1/30	1/60	1/125
5.6	1/4	1/8	1/15	1/30	1/60
8	1/2	1/4	1/8	1/15	1/30
11	1	1/2	1/4	1/8	1/15
16	3	1	1/2	1/4	1/8
22	6	3	1	1/2	1/4
32	15	6	4	2	1/2
45	40	15	9	4	2
64	100	40	20	9	4
100	≫	130	70	30	11
130	≫	265	145	55	20
160	≫	≫	255	100	40
200	≫	≫	≫	180	70
250	≫	≫	≫	≫	130
300	≫	≫	≫	≫	210

SUN – total eclipse: outer corona (10° field)
No filter
$B = 1$

f/	ISO (ASA) 32	ISO (ASA) 64	ISO (ASA) 100	ISO (ASA) 200	ISO (ASA) 400
2	1/8	1/15	1/30	1/60	1/125
2.8	1/4	1/8	1/15	1/30	1/60
4	1/2	1/4	1/8	1/15	1/30
5.6	1	1/2	1/4	1/8	1/15
8	4	2	1/2	1/4	1/8
11	8	3	2	1/2	1/4
16	20	9	5	2	1/2
22	50	19	11	5	2
32	135	50	30	12	5
45	≫	130	70	30	11
64	≫	≫	185	70	30
100	≫	≫	≫	245	95
130	≫	≫	≫	≫	195
160	≫	≫	≫	≫	≫
200	≫	≫	≫	≫	≫
250	≫	≫	≫	≫	≫
300	≫	≫	≫	≫	≫

INDEX

Pages with an illustration are marked with an i and pages with a table are marked with a t. Pages marked "path" show the path of an eclipse across the earth's surface, while a "map" shows additional details, including names of specific locations.